Molecular Mastery

Nanotechnology Fundamentals Unleashed

Geoff Thomas PhD

Copyright © 2024 by Dr Geoff Thomas

All rights reserved. No part of this book may be used or reproduced by any means, graphic, electronic, or mechanical, including photocopying, recording, taping, or by any information storage retrieval system, without the written permission of the publisher except in the case of brief quotations embodied in critical articles and reviews.

Contents

1. Introduction to Nanotechnology 3
2. The Nanoscale World 19
3. Fundamentals of Matter Manipulation 32
4. Nanomaterials: Building Blocks of the Future 45
5. Molecular Nanotechnology: A Revolution Unveiled 62
6. Nanoelectronics and Computing 79
7. Nanomedicine: Healing at the Molecular Level 95
8. Harnessing the Power of the Tiny: Nanotechnology Revolutionizing Energy 111
9. Cultivating Tomorrow: Nanotechnology's Impact on Agriculture 127
10. Environmental Implications of Nanotechnology 143
11. Ethical and Regulatory Dimensions 157
12. Future Prospects and Emerging Trends 171

Dear Reader,

Thank you for choosing to explore the fascinating world of nanotechnology through this book. We hope it provides you with valuable insights and a deeper understanding of the subject. As you delve into the innovative concepts and groundbreaking research presented in this book, we invite you to share your thoughts and reflections by writing a review once you have finished reading.

Your review is incredibly valuable to us and to future readers. Here are some points you might consider including in your review:

1. **Overview and Content**: Provide a brief summary of the book. What are the main topics and themes covered? How does the author approach the subject of nanotechnology?

2. **Clarity and Accessibility**: Discuss the clarity of the writing. Is the book accessible to beginners, or is it more suited for those with prior knowledge in the field? How effectively does the author explain complex concepts?

3. **Strengths and Weaknesses**: Highlight the strengths of the book. What did you find particularly insightful or well-explained? Also, mention any areas where you think the book could be improved.

4. **Practical Applications**: Reflect on any practical applications of the information presented. How might the concepts discussed in the book be applied in real-world scenarios?

5. **Comparison to Other Works**: If you have read other books on nanotechnology, how does this one compare? Does it offer unique perspectives or new information?

6. **Overall Impression**: Summarize your overall impression of the book. Would you recommend it to others interested in nanotechnology? Who would benefit most from reading it?

We encourage you to provide an honest and detailed review. Your insights will not only help others decide whether this book is right for them but also contribute to the ongoing dialogue in the field of nanotechnology.

Thank you for your time and effort in sharing your review. We look forward to reading your thoughts and perspectives.

Best regards,

Geoff Thomas PhD

1. Introduction to Nanotechnology

Introduction

Nanotechnology, often referred to as the science of the small, is a revolutionary field that deals with the manipulation and engineering of matter at the molecular and atomic scale. At this scale, materials exhibit unique properties and behaviours that differ from their bulk counterparts, leading to the development of innovative technologies with applications across various industries. The term "nano" originates from the Greek word "nanos," meaning dwarf. In nanotechnology, materials are manipulated and engineered at dimensions typically ranging from 1 to 100 nanometers. To provide context, one nanometer is equivalent to one-billionth of a meter, or approximately the size of ten atoms arranged in a row. At this scale, the fundamental properties of materials can change dramatically, influencing their optical, electrical, thermal, and mechanical characteristics.

Nanotechnology encompasses a diverse range of disciplines, including physics, chemistry, biology, engineering, and materials science. Its interdisciplinary nature allows researchers to explore and exploit the unique phenomena occurring at the nanoscale in order to develop novel materials, devices, and systems. One of the most remarkable aspects of nanotechnology is its potential to revolutionize various different fields, including medicine, electronics, energy, environmental remediation, and manufacturing. For example, in medicine, nanotechnology enables precise drug delivery, targeted cancer therapy, and advanced diagnostic techniques. In electronics, nanomaterials like graphene and carbon nanotubes offer superior electrical conductivity and mechanical strength, paving the way for smaller, faster, and more efficient electronic devices.

Moreover, nanotechnology plays a crucial role in addressing global challenges such as environmental pollution, water purification, and renewable energy generation. Nanomaterials with unique catalytic properties can facilitate more efficient energy conversion processes, while nanoparticle-based sensors enable rapid detection and monitoring of pollutants in the environment. As nanotechnology continues to advance, its impact on society is expected to grow exponentially. However, along with its vast potential come ethical, safety, and regulatory considerations that must be addressed to ensure responsible development and deployment of nanotechnology-enabled products and technologies. In this chapter, we will delve deeper into the fundamental principles of nanotechnology, explore its applications across various different sectors as well as examine the challenges and opportunities associated with harnessing the power of the nanoscale world.

Pre-Modern Concepts

In the annals of history, ancient civilizations unwittingly engaged with nanoscale phenomena, demonstrating a profound yet unknowing encounter with the nanoworld. One notable instance dates back to the fourth century AD, during the Roman era, where artisans utilized nanoparticles and structures. Although the Romans lacked a formal understanding of the nanoscale, their practical incorporation of these minute structures showcased an early interaction with materials at an atomic and molecular level.

Another intriguing example emerges from ancient Roman glass artefacts, where nanoscale domains were unintentionally created, forming hierarchically assembled photonic crystals. Discovered in archaeological excavations, these glass artefacts revealed unusual nanoscale features that exhibited highly ordered structures, providing a glimpse into the unintentional manipulation of matter at the atomic scale.

Beyond the realms of ancient Rome, diverse cultures displayed inadvertent interactions with nanoscale phenomena. In ancient times, artisans unknowingly manipulated materials, influencing their properties at the nanoscale, as observed in various archaeological findings. These pre-modern encounters with nanoscale effects, while not explicitly understood by ancient societies, underline the inherent human inclination to interact with materials at scales imperceptible to the naked eye. The intersection of history and nanoscience reveals a rich tapestry of early instances where nanoscale phenomena were encountered and manipulated, laying the foundation for the transformative field of nanotechnology we explore today.

The Birth of Nanotechnology

The birth of nanotechnology can be traced back to the visionary insights of physicist Richard Feynman and his seminal 1959 lecture, "There's Plenty of Room at the Bottom." In this groundbreaking address, Feynman articulated a revolutionary concept: the manipulation of matter at the atomic and molecular scale. He envisioned a realm where scientists could control individual atoms, opening up possibilities for crafting materials with unprecedented precision and functionality. Feynman's vision in "There's Plenty of

Room at the Bottom" laid the conceptual groundwork for nanotechnology by challenging the scientific community to think small, urging researchers to explore the vast opportunities at the nanometer length scale. He envisioned a new field of physics and engineering where scientists could manipulate and arrange individual atoms to create custom materials and devices.

This transformative lecture inspired generations of scientists and engineers to delve into the nanoscale world. Feynman's call to action led to the birth of nanotechnology, a field that explores and exploits the unique properties of materials at the atomic and molecular levels. Today, nanotechnology has far-reaching applications in diverse fields, including medicine, electronics, materials science, and energy. Feynman's foresight and passion for exploring the nanoscale laid the foundation for the interdisciplinary field of nanotechnology, where researchers continue to push the boundaries of what is possible, fulfilling his prophetic vision of "plenty of room at the bottom."

Early Developments

Microscopy has a rich history marked by significant developments that have allowed scientists to explore the intricate details of the microscopic world. In the early 17th century, the invention of the compound microscope by Zacharias Janssen and Hans Janssen marked a pivotal moment. This innovation enabled the observation of objects at magnifications previously unattainable, laying the foundation for the study of biological specimens and materials. The 19th century witnessed further advancements with the development of improved lenses and illumination techniques, enhancing the resolution and capabilities of microscopes. The pioneering work of scientists like Robert Hooke and Antonie van Leeuwenhoek contributed to our understanding of cells and microorganisms, opening new realms of exploration.

Fast-forward to the late 20th century, the invention of the scanning tunnelling microscope (STM) in 1981 by Gerd Binnig and Heinrich Rohrer revolutionized microscopy. Unlike traditional optical microscopes, the STM operates by scanning a sharp metal tip over a surface, detecting the tunnelling current between the tip and the sample. This breakthrough allowed scientists to visualize and manipulate matter at the atomic level, ushering in the era of nanoscale observation. The STM played a crucial role in enabling nanoscale observation by offering unprecedented resolution and precision. Its ability to image individual atoms and molecules paved the way for advancements in nanotechnology, materials science, and condensed matter physics. The STM exemplifies the continual evolution of microscopy, providing scientists with powerful tools to explore and manipulate the nanoscale world.

Emergence as a Field

The emergence of nanotechnology as a distinct field can be traced back to the late 20th century, marked by advancements in manipulating matter at the nanoscale. The term "nanotechnology" was popularized by physicist Richard Feynman in his famous 1959 lecture, "There's Plenty of Room at the Bottom," where he envisioned the possibility of manipulating individual atoms and molecules. However, it wasn't until the 1980s and 1990s that significant breakthroughs, such as the invention of the scanning tunnelling microscope (STM), allowed scientists to directly observe and manipulate materials at the nanoscale. Nanotechnology's interdisciplinary nature became apparent as researchers from various scientific disciplines converged to explore its potential. Nanoscience, the study of phenomena at the nanoscale, began to formalize as an interdisciplinary field encompassing physics, chemistry, materials science, and biology. This convergence of diverse scientific domains led to the recognition of nanoscience and nanotechnology as distinct and interrelated disciplines. Formal academic programs and institutions dedicated to nanoscience and nanotechnology further solidified their status. For example, master's programs, like the Master in Nanoscience and Nanotechnology, have been established to provide comprehensive interdisciplinary education in this emerging field. Nanotechnology has revolutionized industries such as electronics, medicine, energy, and materials science, underlining the importance of recognizing and formalizing it as a field that transcends traditional disciplinary boundaries.

Key Principles

Nanotechnology relies on several fundamental principles that govern the behaviour of matter at the nanoscale, enabling the design and manipulation of materials with unprecedented precision. These core principles include:

1. Quantum Mechanics:

At the heart of nanotechnology lies the principles of quantum mechanics, a branch of physics that explores the behaviour of particles at the atomic and subatomic levels. Quantum effects become prominent at the nanoscale, influencing the electronic, optical, and magnetic properties of materials.

2. Surface Science:

Understanding and manipulating the properties of surfaces is crucial in nanotechnology. As the surface-to-volume ratio increases significantly at the nanoscale, surface interactions play a dominant role in the behaviour of nanomaterials. Surface science principles are applied to engineer and modify surface properties for specific applications.

3. Self-Assembly:

Self-assembly is a fundamental concept in nanotechnology, wherein nanoscale structures spontaneously organize into ordered patterns or arrangements. This process mimics natural phenomena and is advantageous for creating complex structures with minimal external intervention. Self-assembly is harnessed for nanofabrication and the creation of nanomaterials with unique properties.

These principles collectively form the foundation of nanotechnology, allowing scientists and engineers to manipulate matter at the atomic and molecular levels. The interdisciplinary nature of nanotechnology draws from physics, chemistry, materials science, and engineering to harness these principles for a wide range of applications, from medicine and electronics to energy and materials development.

Milestones and Achievements

Nanotechnology has witnessed remarkable milestones and achievements, marking its evolution from theoretical concepts to transformative applications. Key breakthroughs include:

1. Pre-Modern Examples:

Nanotechnology's roots extend to ancient times with the crafting of nanomaterials like Damascus steel, showcasing early nanoscale engineering without a scientific understanding.

2. 1981 - Scanning Tunnelling Microscope (STM):

Gerd Binnig and Heinrich Rohrer's invention of the STM allowed visualization of individual atoms, laying the foundation for nanoscale observation.

3. 1985 - Fullerenes (Buckyballs):

The discovery of fullerenes by Robert Curl, Sir Harold Kroto, and Richard Smalley opened new avenues. These soccer ball-shaped molecules introduced novel nanomaterials.

4. 1991 - Carbon Nanotubes:

Sumio Iijima's identification of carbon nanotubes revolutionized material science. These structures exhibited remarkable strength and conductivity.

5. 1999 - Quantum Dots for Biological Imaging:

Developed by Moungi Bawendi and colleagues, quantum dots emerged as vital tools in biological imaging, enabling precise visualization at the cellular level.

6. 2004 - Nobel Prize for STM and AFM:

The Nobel Prize in Physics was awarded to the inventors of the Scanning Tunnelling Microscope (STM) and Atomic Force Microscope (AFM), recognizing their impact on nanotechnology.

7. 2010 - Graphene Isolation:

The isolation of graphene by Andre Geim and Konstantin Novoselov marked a breakthrough, leading to the exploration of its unique properties.

8. CRISPR-Cas9 Nanotechnology Applications:

Integrating CRISPR-Cas9 with nanotechnology offers precise gene editing, holding promise for medical advancements.

These achievements illustrate nanotechnology's evolution from fundamental discoveries to practical applications, impacting diverse fields like materials science, medicine, and electronics.

National Nanotechnology Initiative (NNI)

The National Nanotechnology Initiative (NNI) plays a pivotal role in coordinating and advancing nanotechnology research and development efforts across various U.S. government agencies. Established in 2000, the NNI emerged in response to the growing recognition of nanotechnology's potential impact on diverse sectors, including medicine, electronics, and materials science. Formed under the auspices of the National Science and Technology Council, the NNI brings together multiple federal agencies, such as the National Institutes of Health (NIH), the National Aeronautics and Space Administration (NASA), and the National Institute of Standards and Technology (NIST), to collaboratively drive nanotechnology initiatives. The initiative operates with the vision of harnessing nanoscale science and engineering to address critical challenges and foster innovation.

The NNI focuses on several key areas, including fundamental research, infrastructure development and addressing societal implications and ethical considerations of nanotechnology. By fostering interdisciplinary collaborations, the NNI aims to accelerate discoveries and promote the responsible development of nanotechnology. Governmental involvement with nanotechnology through the NNI is manifested in strategic investments, collaborative programs, and the establishment of research centres. The NNI's role in shaping nanotechnology research and development is evident in its efforts to propel groundbreaking discoveries, facilitate knowledge-sharing, and guide the ethical implementation of nanotechnological advancements. The NNI stands as a testament to the U.S. government's commitment to advancing nanotechnology, providing a framework that promotes collaboration, innovation, and responsible development in this transformative field.

Global Perspectives

Global perspectives and international collaborations are paramount in advancing nanotechnology, a field that thrives on interdisciplinary cooperation and shared scientific endeavours. The collaborative landscape in nanotechnology is marked by extensive international efforts, fostering innovation, knowledge exchange, and the pursuit of breakthroughs. International collaboration in nanotechnology is evident in various aspects, including research and development initiatives, joint projects, and collaborative publications. Scientists and researchers from different nations collaborate to pool their expertise and resources, driving progress in nanoscale science and engineering. This collaborative approach accelerates the pace of discovery, ensuring that advancements benefit from a diverse range of perspectives.

Nanotechnology patents serve as a tangible manifestation of international collaboration, reflecting joint efforts to develop novel technologies. Patent network analyses highlight the interconnectedness of global nano research and development, emphasizing the significance of cross-border partnerships in advancing the field. Conferences and exhibitions on a global scale provide platforms for researchers to showcase their innovative findings and foster international collaborations. These events serve as hubs for networking, enabling scientists to form partnerships that transcend geographical boundaries.

Furthermore, nanotechnology plays a crucial role in addressing global challenges such as sustainability. International collaborations leverage nanotechnology for sustainable development, emphasizing its role in environmental conservation, energy efficiency, and addressing societal needs. In conclusion, the global perspective on nanotechnology is shaped by collaborative efforts that transcend national borders. The exchange of knowledge, resources, and ideas on an international scale is instrumental in propelling nanotechnology forward, ensuring that the benefits of this transformative field are realized globally.

Nanotechnology Today

Nanotechnology today occupies a dynamic landscape with a broad spectrum of applications, presenting both opportunities and challenges. In contemporary industries, nanotechnology finds widespread use, notably in medicine, electronics, materials science, and environmental monitoring.

Applications:

1. Medicine:

Nanotechnology has revolutionized drug delivery systems, enabling targeted treatments with minimized side effects. Nanoparticles are used for imaging, diagnostics and therapy in various medical applications.

2. Electronics:

The semiconductor industry benefits from nanoscale innovations, enhancing the performance of electronic devices. Nanomaterials contribute to the development of faster and more efficient electronic components.

3. Materials Science:

Nanocomposites and nanomaterials have enhanced the mechanical, thermal, and electrical properties of materials, leading to the creation of advanced materials with diverse applications.

4. Environmental Monitoring:

Nanosensors play a crucial role in environmental monitoring, detecting pollutants and contaminants with high sensitivity.

Challenges:
1. Regulatory Challenges:

The regulatory landscape faces challenges related to the lack of global standardization in nomenclature, test methods and characterization of nanomaterials.

2. Health Concerns:

The impact of nanoparticles on health is a subject of ongoing research, addressing concerns about their potential effects on human health.

3. Ethical Considerations:

As nanotechnology advances, ethical considerations surrounding its applications, especially in medicine and surveillance, require careful examination.

Ongoing Research:
1. Biomedical Advancements:

Ongoing research explores nanotechnology's potential in personalized medicine, regenerative medicine, and nanorobotics for targeted therapies.

2. Sustainability:

Nanotechnology is actively researched for sustainable practices, including efficient energy storage, water purification, and eco-friendly manufacturing processes.

The current nanotechnology landscape reflects a thriving field with diverse applications, though challenges and ongoing research underscore the need for responsible development and continued exploration.

Future Directions

The future of nanotechnology promises transformative advancements across diverse industries, fostering innovation and addressing complex challenges.

1. Medicine:

Anticipated advancements in nanomedicine include personalized drug delivery systems, targeted cancer therapies, and diagnostic nanosensors. Nanoparticles designed to navigate the human body at the molecular level could revolutionize treatment efficacy and minimize side effects.

2. Electronics:

Nanotechnology is poised to revolutionize the electronics industry with the development of smaller, more powerful, and energy-efficient devices. Quantum computing, enabled by nanoscale components, holds the potential to surpass classical computing capabilities, leading to unprecedented advancements in information processing.

3. Materials Science:

Nanomaterials are expected to revolutionize materials science, resulting in the creation of stronger, lighter, and more versatile materials. Applications range from advanced aerospace components to durable construction materials.

4. Environmental Applications:

Nanotechnology may play a crucial role in addressing environmental challenges. Nanoscale sensors for real-time pollution monitoring, advanced nanomaterials for water purification, and efficient energy storage solutions are among the speculated future developments.

As nanotechnology progresses, ethical considerations, safety protocols, and regulatory frameworks will be essential to guide responsible innovation. The continuous collaboration between scientists, engineers, and policymakers is crucial to harness the full potential of nanotechnology while mitigating potential risks. The future of nanotechnology holds great promise, impacting medicine, electronics, materials science and environmental sustainability. As research and development in nanotechnology advance, society can anticipate groundbreaking solutions to some of its most pressing challenges.

Conclusion

The evolution of nanotechnology has been a remarkable journey, revolutionizing industries and reshaping our approach to science and technology. From its conceptualization to the present day, nanotechnology has witnessed a transformative trajectory, propelling us into a realm where the manipulation of matter at the nanoscale has become a reality. Historically, nanotechnology's roots can be traced back to visionary thinkers who laid the theoretical groundwork. Over time, researchers and scientists have unravelled the mysteries at the nanoscale, leading to breakthroughs that have fundamentally altered various sectors.

The National Nanotechnology Initiative (NNI) has played a pivotal role in guiding nanotechnology's development, emphasizing responsible and transformative advancements. Investments and decisions made under the NNI have steered the field towards a future marked by innovation and societal benefits. Recent advances in nanotechnology showcase its limitless scope, with applications ranging from medicine and electronics to materials science and environmental solutions. Nanotechnology has introduced novel materials, enhanced drug delivery systems, and contributed to the development of more efficient and sustainable technologies.

In conclusion, nanotechnology's historical journey reflects a relentless pursuit of understanding and manipulating matter at the smallest scale. The transformative impact of nanotechnology is evident in its multifaceted applications and the promise it holds for addressing global challenges. As we stand on the brink of a nanotechnological era, collaboration, ethical considerations, and continued research will be essential in harnessing its full potential for the betterment of society.

2. The Nanoscale World

Mastering the Molecular Realm

Nanotechnology, a revolutionary field at the intersection of science and technology, delves into the manipulation of materials at the molecular and atomic scale. At its core, nanotechnology operates on the premise of understanding, controlling, and engineering matter at dimensions smaller than 100 nanometers, where the properties of materials exhibit unique and often unprecedented characteristics. This chapter unravels the intricate world of nanotechnology, offering a comprehensive exploration of its principles, applications, and the profound implications it holds for various industries.

Defining Nanotechnology

Nanotechnology encompasses two fundamental components: nanoscience and nanotechnology itself. Nanoscience involves the study of phenomena at the nanoscale, examining materials at atomic, molecular and macromolecular levels where properties diverge significantly from the macroscopic scale. On the other hand, nanotechnology, the applied aspect, empowers scientists and engineers to manipulate matter at these tiny scales, providing unprecedented control over materials and their functionalities.

Significance of Molecular Mastery

Understanding and controlling matter at the molecular and atomic scale heralds a new era of possibilities and advancements. The significance lies in the unique properties that emerge at the nanoscale. Materials exhibit altered physical, chemical, and biological behaviours, opening avenues for innovation across diverse domains. Nanotechnology has already shown its prowess in fields such as medicine, electronics, and materials science, where tailored materials and devices lead to enhanced performance, increased efficiency, and novel functionalities. The ability to work at the atomic and molecular levels provides unparalleled precision. Researchers can observe, move, and manipulate individual atoms, paving the way for the creation of nanoscale structures with meticulous detail. This precision not only transforms the manufacturing processes but also enables the design of materials with tailored properties, contributing to the development of advanced technologies and cutting-edge applications.

Nanotechnology serves as a gateway to a realm where the rules of classical physics begin to blur, and quantum effects come to the forefront. Harnessing these effects allows for the creation of materials and devices with unprecedented functionalities, promising breakthroughs that were once deemed impossible. As we embark on this journey into the molecular realm, the chapters that follow will delve deeper into the diverse applications, ethical considerations, and the limitless potential that nanotechnology holds for shaping the future of science and technology.

Unravelling the Tapestry of Nanotechnology

The historical development of nanotechnology is a captivating journey that intertwines ancient practices, visionary pioneers, and the convergence of multiple scientific disciplines. Early pioneers in nanotechnology, Richard Smalley, Robert Curl, and Sir Harold Kroto, were awarded the Nobel Prize in Chemistry in 1996 for their discovery of fullerenes. These carbon molecules, also known as buckyballs, represented a new class of nanomaterials with unique properties. The discovery of fullerenes showcased the potential for creating novel materials at the nanoscale and fuelled the burgeoning interest in nanotechnology. Acknowledging the interdisciplinary nature of nanotechnology, it was recognized as a convergence of physics, materials science, and biology. Nanoscience emerged as a distinct discipline studying phenomena and manipulating materials at atomic, molecular, and macromolecular scales. This convergence allowed researchers to integrate principles from various scientific domains, creating a holistic understanding of matter at the nanoscale.

Nanoscience's recognition as a convergence of physics, materials science, and biology marked a paradigm shift in scientific exploration. The interdisciplinary approach facilitated breakthroughs in diverse fields, including medicine, electronics, and materials science. Researchers began to explore and harness the unique properties of materials at the nanoscale, leading to innovations with profound implications for technology and medicine. The historical development of nanotechnology reflects a rich tapestry woven with ancient practices, visionary pioneers, and the convergence of scientific disciplines. From ancient civilizations' inadvertent engagement in nanoscale manipulation to the explicit recognition of nanotechnology as a field, the journey has been transformative. Early pioneers and key milestones have propelled nanotechnology into a multidisciplinary realm with the potential to reshape the future of science and technology.

Navigating the Nanoscale Landscape

Nanotechnology, a groundbreaking field at the intersection of physics, materials science, and biology, operates on the nanoscale — a realm where materials exhibit unique properties due to their dimensions falling within the range of 1 to 100 nanometers. This scale is extraordinary; representing a billionth of a meter, and it is at this minute level that the distinct characteristics of nanotechnology come to life. At the nanoscale, materials undergo a fundamental shift in their properties compared to their macroscopic counterparts. One defining feature is the increased surface area-to-volume ratio, offering a platform for heightened reactivity and novel behaviours. This phenomenon becomes particularly pronounced in nanomaterials, where the increased surface area allows for enhanced interactions with surrounding elements.

Quantum effects dominate the nanoscale, introducing an intriguing aspect of nanotechnology. Unlike classical physics, quantum mechanics governs the behaviour of particles at this scale. Materials can exhibit quantum confinement, where the movement of electrons becomes confined in three dimensions, leading to quantized energy levels. This phenomenon contributes to the unique optical, electronic, and magnetic properties observed in nanomaterials. Another striking aspect of nanotechnology is the ability to engineer materials with precision at the atomic and molecular levels. Researchers can manipulate matter atom by atom, paving the way for the creation of tailored structures with desired functionalities. This precision craftsmanship is exemplified by the invention of the scanning tunnelling microscope (STM), which allows scientists to visualize and manipulate individual atoms. The

STM opened avenues for nanoscale exploration and manipulation, shaping the trajectory of nanotechnology research.

Nanotechnology is not confined to one specific field; rather, it permeates across various disciplines. In medicine, nanoparticles are employed for targeted drug delivery, taking advantage of their unique properties to enhance therapeutic efficacy while minimizing side effects. In electronics, nanoscale components enable the creation of smaller, more efficient devices with improved performance. The field also intersects with materials science, introducing nanomaterials with enhanced strength, flexibility, and conductivity. The phenomenon of self-assembly is another hallmark of nanotechnology. At the nanoscale, materials have an intrinsic tendency to organize themselves into ordered structures. This spontaneous self-assembly allows for the creation of intricate nanoscale architectures without external intervention, a feature harnessed in the fabrication of nanodevices and nanomaterials.

The fundamentals of nanotechnology lie in the exploration of the nanoscale, where materials exhibit distinct characteristics and behaviours. The increased surface area-to-volume ratio, quantum effects, precision engineering at the atomic level, and self-assembly are key elements contributing to the uniqueness of nanotechnology. As researchers delve deeper into the nanoscale landscape, the field continues to unveil unprecedented possibilities, promising transformative applications across diverse scientific domains.

The Crucial Role of Nanoscale Science and Technology

Nanoscale science, commonly known as nanoscience, stands at the forefront of scientific innovation, playing a pivotal role in shaping the future of various domains. Operating at the nanometer scale, where materials exhibit unique properties, nanoscience has become synonymous with groundbreaking advancements in technology and research.

Precision Engineering with Nanotechnology

Nanotechnology, a direct offspring of nanoscience, involves the manipulation and application of materials at the nanoscale. One of the key aspects is the ability to engineer nanomaterials with tailored properties, opening avenues for advancements across various scientific domains. For instance, in medicine, nanotechnology facilitates targeted drug delivery, enhancing treatment efficacy while minimizing side effects. The precision offered by nanotechnology has revolutionized diagnostics and therapeutic interventions.

Materials Science and Nanotechnology

Nanoscale science significantly contributes to materials science through the creation of nanomaterials. These materials possess unique characteristics due to their size, leading to enhanced strength, conductivity, and reactivity. In the realm of electronics and computing, nanotechnology enables the development of smaller and more efficient devices. Continuous advancements in nanoscale engineering have been crucial in driving the semiconductor industry towards more powerful and compact electronic components.

Biomedical Applications

The biomedical field has witnessed substantial breakthroughs owing to nanotechnology. Diagnostic techniques with unprecedented sensitivity, such as nanoparticle-based imaging, are now possible.

Moreover, nanotechnology has paved the way for innovative approaches in cancer treatment, showcasing its potential to revolutionize healthcare.

Environmental Impact and Energy Solutions

Nanoscience contributes to addressing environmental challenges and advancing energy solutions. Nanomaterials are employed in catalysts for clean energy production, while energy storage devices benefit from enhanced performance due to nanoscale engineering.

Nanoscale science, encapsulated in the realm of nanotechnology, is a driving force behind advancements in medicine, electronics, materials science, and environmental solutions. Its interdisciplinary nature and precision engineering capabilities position nanoscience as a cornerstone for continued innovation across diverse scientific domains.

Empirical Exploration of Nanoscience

Nanoscience, defined as the understanding and control of matter at the nanoscale (0.1 to 100 nanometers) has been subject to extensive empirical exploration, unravelling unique phenomena that arise at these dimensions. Empirical studies play a crucial role in investigating the principles governing nanomaterials and their distinctive properties.

Principles and Phenomena at the Nanoscale

Empirical exploration in nanoscience involves studying materials and systems within the nanometer range to comprehend their behaviour. A notable empirical study showcased in the literature is the review by Bayda et al. (2020), which illustrates the progress and main principles of nanoscience and nanotechnology. This comprehensive exploration covers both pre-modern and modern aspects, emphasizing the evolution of our understanding and control of matter at the nanoscale.

Understanding and Controlling Matter

Empirical studies focus on understanding and controlling matter at the nanoscale, where unique properties emerge. Nanotechnology, as an offspring of nanoscience, thrives on the ability to engineer materials with precision. An empirical exploration reported on ResearchGate delves into the definition of nanotechnology and emphasizes the understanding and control of matter at dimensions between 0.1 to 100 nanometers. This dimensionality unlocks distinctive phenomena, enabling advancements in various scientific domains.

Characterization Techniques for Nanoparticles

Empirical studies also encompass the development and application of characterization techniques for nanoparticles. Mourdikoudis et al. (2018) present a detailed exploration of techniques used to characterize the size, crystal structure, elemental composition, and physical properties of nanoparticles. These techniques are essential for gaining insights into the unique characteristics of nanomaterials, contributing to the empirical foundation of nanoscience.

The Nanoscale and Unique Phenomena

Empirical exploration at the nanoscale is essential for uncovering unexpected physical and chemical characteristics. Nanotechnology and nanomaterials exhibit properties distinct from bulk materials, as highlighted in studies on ScienceDirect. This exploration emphasizes the importance of dimensions ranging from 1 to 100 nanometers, where materials showcase unique phenomena that have profound implications across various scientific domains.

The empirical exploration of nanoscience through various studies has been instrumental in advancing our understanding of the principles and phenomena at the nanoscale. These studies pave the way for harnessing the unique properties of nanomaterials, driving innovations in nanotechnology and contributing to diverse scientific disciplines.

Unravelling Philosophical Perspectives on Nanoscience

Nanoscience, at the intersection of science and philosophy, prompts diverse perspectives from various disciplines, shedding light on its philosophical underpinnings. Different disciplinary viewpoints, notably from chemistry and molecular biology, contribute to the evolving narrative of nanotechnology.

1. Interdisciplinary Nature

Nanoscience's emergence can be viewed as an "engineering way of being in science," a term coined by philosopher Alfred Nordmann. This characterizes nanoscientific research as inherently interdisciplinary, drawing from multiple domains such as chemistry, physics, and engineering. This interdisciplinary nature inherently involves philosophical considerations about the nature of scientific knowledge and collaboration.

2. Chemistry's Contribution

Chemistry plays a pivotal role in the philosophical landscape of nanoscience. The synthesis and manipulation of nanomaterials often fall within the realm of chemistry. Researchers in this field explore philosophical questions related to the ethical implications of manipulating matter at the nanoscale, the nature of chemical interactions and the boundaries of human intervention at the molecular level.

3. Molecular Biology's Influence

The progression of nanotechnology from molecular biology is noteworthy. Molecular biology has historically dominated discussions in biology, and its influence extends into nanoscience. Philosophical perspectives from molecular biology highlight concerns and considerations about synthetic biology and the design of synthetic organisms at the nanoscale. This transition reflects the natural evolution of scientific inquiry and the integration of molecular-level insights into nanotechnology.

4. Ethical and Social Dimensions

Philosophical considerations extend beyond scientific disciplines to encompass ethical and social dimensions. A review on the philosophy and social aspects of nanotechnology emphasizes the importance of exploring these perspectives to better understand the implications of nanotechnology. This includes ethical dilemmas, societal impacts and the need for responsible development and deployment of nanotechnological advancements.

The philosophical perspectives of nanoscience are deeply intertwined with its interdisciplinary nature, drawing insights from chemistry, molecular biology, and beyond. As nanotechnology continues to advance, these perspectives contribute to a rich tapestry of ethical, societal, and scientific reflections.

The Future Implications of Nanoscience

Nanoscience, the study and manipulation of matter at the molecular and atomic scale, holds immense promise for the future, ushering in revolutionary advancements across various industries. A deep understanding of these minuscule dimensions opens a gateway to unprecedented applications, paving the way for transformative impacts.

1. Nanomedicine: Revolutionizing Healthcare

Nanotechnology is poised to revolutionize healthcare with groundbreaking applications in nanomedicine. Precise drug delivery to targeted cells, enabled by nanoparticles, reduces side effects and enhances therapeutic efficacy. The development of nanoscale diagnostic tools promises earlier and more accurate disease detection, fundamentally altering the landscape of medical practice.

2. Materials Science and Construction: Stronger, Lighter, Smarter

In materials science and construction, nanoscience offers the potential to create materials with enhanced properties. Manipulating materials at the atomic scale results in stronger, lighter, and more durable substances. From advanced building materials to innovative solutions in manufacturing and packaging, nanotechnology is redefining the possibilities in materials engineering.

3. Electronics and Device Engineering: Smaller, Faster, Energy-Efficient

The electronics industry is on the cusp of a transformative era, thanks to nanoscience. Miniaturization at the atomic scale allows for the creation of more efficient and powerful electronic components. This could lead to the development of smaller, faster, and more energy-efficient devices, impacting areas from wearables to computing systems.

4. Environmental Impact and Energy: Sustainability Redefined

Nanoscience contributes to addressing environmental challenges with applications in water purification, pollution remediation, and renewable energy. Nanomaterials enable efficient water purification, while innovations in nanotechnology may reshape the energy landscape through advancements in solar energy harnessing, offering sustainable alternatives to conventional practices.

5. Cyber and Robotics: Enhancing Capabilities

Beyond the tangible, nanotechnology impacts the realms of cyber and robotics. Nanoscale components enhance the capabilities of sensors and computing systems, influencing the development of more sophisticated robotics and cybersecurity solutions. This intersection highlights the interconnectedness of nanoscience with cutting-edge technologies, setting the stage for a new era of innovation.

6. Broad Industry Applications: Diverse Transformations

The transformative impact of nanoscience spans across diverse industries, from medicine to electronics and environmental sustainability. Nanotechnology's versatility positions it as a driving force behind a wave of innovations, shaping the future trajectory of global industries. As our understanding of the molecular and atomic scale deepens, the applications and advancements arising from nanoscience are poised to redefine the way we live, work, and interact with the world.

The future implications of nanoscience are vast and transformative. The ability to manipulate matter at the molecular and atomic scale unlocks a multitude of applications, offering solutions to pressing global challenges and revolutionizing industries in ways previously unimaginable.

3. Fundamentals of Matter Manipulation

The Fundamentals of Nanotechnology

Nanotechnology, the science of manipulating matter at the molecular and atomic levels, marks a revolutionary leap in scientific exploration. This field encapsulates a multidisciplinary approach, intertwining nanoscale science, engineering, and technology. At its core, nanotechnology involves the imaging, measuring, modelling, and manipulation of matter with dimensions typically ranging from 1 to 100 nanometers.

Matter Manipulation at the Molecular Level

Matter manipulation at the molecular level within nanotechnology is akin to playing with building blocks on an unimaginably small scale. Scientists and engineers leverage this capability to meticulously arrange atoms and molecules, creating materials and structures with unparalleled precision. At the nanoscale, quantum effects become predominant, leading to unique properties that diverge from those observed in macroscopic materials. Nanotechnology harnesses techniques such as scanning tunnelling microscopy and atomic force microscopy, enabling researchers to visualize and manipulate individual atoms. The ability to control matter at such minute scales opens the door to a realm of possibilities, allowing scientists to engineer materials with tailored properties and functionalities.

Significance and Applications

The significance of nanotechnology reverberates across diverse scientific domains, ushering in transformative applications with far-reaching implications.

1. Medicine and Healthcare

Nanomedicine exploits nanotechnology for targeted drug delivery, imaging, and diagnostics. Nanoparticles can navigate biological barriers, enhancing treatment efficacy and reducing side effects.

2. Electronics and Computing

In electronics, nanotechnology enables the development of smaller, faster, and more efficient devices. Quantum dots and nanowires facilitate the creation of nanoscale transistors and memory devices.

3. Energy

Nanomaterials play a pivotal role in energy storage, conversion, and harvesting. Nanotechnology contributes to the development of advanced batteries, solar cells, and fuel cells.

4. Materials Science

Nanocomposites, materials composed of nanoscale components, exhibit enhanced mechanical, electrical, and thermal properties. This has applications in manufacturing, construction, and aerospace.

5. Environmental Remediation

Nanotechnology offers solutions for environmental challenges. Nanoparticles can be employed for water purification, pollution detection, and remediation processes.

The fundamentals of nanotechnology empower scientists to manipulate matter at unprecedented scales, unlocking a plethora of applications across scientific disciplines. This introduction provides a glimpse into the intricate world of nanotechnology, where the manipulation of matter at the molecular level holds the key to addressing pressing challenges and fostering innovation.

Tools of the Trade
1. Scanning Tunnelling Microscopy (STM)

Scanning Tunnelling Microscopy, or STM, stands as a pioneering technique in the realm of nanoscience. Operating on the principles of quantum tunnelling, STM employs a sharp metallic tip positioned near a conductive sample. As the tip hovers extremely close to the surface, electrons quantum mechanically tunnel between the tip and the sample. This tunnelling current is highly sensitive to the tip-sample separation, allowing for the precise mapping of atomic structures. STM's unparalleled spatial resolution, down to the atomic scale, makes it an invaluable tool for high-precision molecular imaging. Researchers harness this capability to explore surfaces with atomic details, unravelling the mysteries of materials and molecules at an unprecedented level.

2. Optical Tweezers

Optical tweezers utilize the momentum of light to manipulate microscopic objects, representing a groundbreaking advancement in the field of optics. By employing focused laser beams, researchers can trap and manipulate particles, ranging from nanometre to micrometre scales. The principles behind optical tweezers rely on the gradient force created by the light intensity gradient. This force acts as microscopic "tweezers," allowing scientists to trap and precisely control the position of particles. In the biological and chemical sciences, optical tweezers find diverse applications. They enable the delicate manipulation of cells, organelles, and even individual biomolecules. Researchers use optical tweezers to apply controlled forces, study molecular interactions and to explore the biomechanics of living organisms.

3. Atomic Force Microscopy (AFM)

Atomic Force Microscopy, or AFM, extends the capabilities of traditional microscopy by probing and manipulating surfaces at the atomic level. AFM utilizes a sharp tip attached to a cantilever to scan the surface of a sample. As the tip interacts with the sample's surface, the deflection of the cantilever is measured, providing information about the topography and mechanical properties of the material. AFM excels in nanoscale topography control, enabling researchers to visualize and manipulate structures with remarkable precision. In addition to imaging, AFM is employed for various applications, including nanolithography, surface modification, and the study of biological macromolecules. Its versatility makes AFM an indispensable tool for both material scientists and biologists, facilitating advancements in diverse fields. These tools of the trade — Scanning Tunnelling Microscopy, Optical Tweezers, and Atomic Force Microscopy — unlock the secrets of the microscopic world. Each contributes to our understanding of matter at different scales, offering unprecedented capabilities for imaging, manipulation, and exploration.

Techniques for Molecular Manipulation
1. Electrically Controlled Manipulation

Electrically controlled manipulation stands at the forefront of molecular precision control. This technique involves the application of electrical stimuli to molecules, harnessing the power of electrostatic forces for precise manipulation. By strategically applying electrical fields, researchers can govern the orientation, position, and even conformation of molecules with unprecedented accuracy. This approach offers a versatile toolkit for engineering molecular structures, paving the way for advancements in nanoelectronics, materials science, and molecular engineering. The ability to fine-tune molecular arrangements through electrical means holds immense potential for designing novel materials with tailored properties.

2. Lithographic Imaging Techniques

Lithographic imaging techniques play a pivotal role in crafting nanoscopic features at the atomic and molecular levels. This method involves the use of lithography, a process akin to traditional printing, but on a minuscule scale. Researchers can precisely control the patterning of materials by selectively exposing them to specific wavelengths of light. This controlled exposure enables the formation of intricate nanostructures, essential for engineering materials with desired properties. Lithographic imaging techniques find applications in diverse fields, from semiconductor manufacturing to the creation of nanoscale devices. The ability to engineer materials at such fine scales opens avenues for innovation in electronics, medicine, and materials science.

3. Optical Manipulation Techniques

Optical manipulation techniques harness the power of light for non-invasive control of molecular structures. Using techniques such as optical tweezers, researchers can trap, move, and manipulate microscopic objects with high precision. Optical tweezers utilize focused laser beams to exert forces on particles, allowing researchers to perform delicate manipulations without physical contact. In the realm of molecular biology, these techniques enable the study of cellular processes and the manipulation of individual biomolecules. The non-invasive nature of optical manipulation techniques is particularly advantageous for studying living cells and biological entities without causing damage. This approach opens avenues for advancements in biophysics, pharmacology, and nanomedicine.

These techniques for molecular manipulation—electrically controlled manipulation, lithographic imaging techniques, and optical manipulation techniques—usher in a new era of precision engineering at the molecular scale. Each method contributes uniquely to the manipulation and engineering of molecular structures, offering unprecedented possibilities for scientific and technological advancements.

Molecular Nanotechnology: Shaping the Future at the Molecular Scale
Engineering Nanoscale Wonders

Molecular nanotechnology (MNT) emerges as a groundbreaking field, aiming to engineer and manipulate materials at the molecular scale, unlocking unparalleled possibilities. At its core, MNT operates on the fundamental principles of manipulating individual molecules to construct precise, atomically defined structures. This approach, often referred to as mechanosynthesis, allows researchers to build materials with atomic precision, ushering in a new era of precision engineering. Working with nanoscale structures involves leveraging molecular machines and nanodevices that operate on the nanometre scale. These nanoscale structures exhibit unique properties, often vastly different from their macroscopic counterparts. The principles of MNT enable the design and assembly of nanoscale components, paving the way for the development of advanced materials, sensors, and devices.

The Future Implications: Advancements and Transformative Potential

The future implications of molecular nanotechnology are vast and transformative, promising advancements that span across various industries. In medicine, MNT holds the potential for targeted drug delivery systems, personalized medicine, and innovative treatments at the molecular level. The ability to engineer nanoscale devices for medical applications opens avenues for precise diagnostics and therapeutics. In the realm of electronics, MNT could revolutionize the manufacturing of nanoscale components, leading to smaller and more efficient devices. This transformative potential extends to energy production, where nanoscale materials could enhance the efficiency of solar cells and energy storage devices. MNT's cross-industry impact extends to materials science, where the creation of novel materials with tailored properties becomes a reality.

The development of lightweight yet incredibly strong materials has implications for aerospace and manufacturing. Furthermore, environmental applications could benefit from nanoscale technologies, offering solutions for pollution remediation and sustainable energy production. Molecular nanotechnology therefore emerges as a scientific frontier with the power to redefine industries and revolutionize our technological landscape. As researchers delve deeper into the principles of molecular manipulation and nanoscale engineering, the future implications of MNT become increasingly promising. The transformative potential of MNT not only opens doors to technological advancements but also presents a paradigm shift in the way we approach challenges across diverse fields.

Nanoscale Manipulation Unleashing Real-World Potential Nanoscale Insights Leading to Breakthroughs

In the realm of nanotechnology, case studies provide valuable insights into the practical applications of nanoscale manipulation. These studies delve into the innovative techniques that researchers employ to manipulate matter at the nano level, showcasing the transformative impact on various industries. One notable case study involves the application of deep reinforcement learning (DRL) in micro-electro-mechanical systems (MEMS) and nanotechnology. DRL has proven instrumental in enhancing the capabilities of MEMS, showcasing how cutting-edge technologies, like artificial intelligence, can be harnessed for nanoscale manipulation. Optical manipulation stands out as another key area explored in case studies, ranging from the microscale to the nanoscale. Optical traps, one of the preferred methods, allow the manipulation of objects in microscopic systems, finding applications in physics and beyond.

Understanding these optical manipulation techniques is crucial for unlocking new possibilities in nanoscale assembly and manipulation.

Real-World Applications of Nano-Optics

Nano-optics, a field focused on manipulating light at dimensions comparable to the diffraction limit, has been a subject of intense study. Case studies reveal the practical applications of nano-optics in diverse fields. This includes advancements in manipulating light for small-scale devices, opening avenues for innovations in communication, sensing, and imaging technologies. Additionally, case studies explore the impact of engineered nanoparticles in daily life. The review discusses the various entry routes of nanoparticles and their applications, shedding light on the intersection of nanotechnology and everyday products. Understanding the implications of nanoparticles in real-world scenarios is essential for managing their toxicity and harnessing their benefits.

Progress in Nanotechnology across Industries

Professor Wenlong Cheng's case studies exemplify the progress in nanotechnology with a focus on creating materials with exotic properties and real-world applications. The manipulation and control of materials at the nanoscale offer unprecedented opportunities in material science, with potential impacts on various industries. In a broader context, recent advances in nanotechnology serve as case studies themselves, showcasing significant progress across various fields. These advances involve the manipulation and control of nanoscale structures, emphasizing the cross-industry impact of nanotechnology on fields such as medicine, electronics, and materials science. Case studies in nanotechnology therefore provide a window into the practical applications of nanoscale manipulation, unveiling a world of possibilities for innovation and advancement across diverse industries.

Challenges and Ethical Considerations in Molecular Nanotechnology
Technical Challenges and Emerging Solutions

Molecular nanotechnology, while promising transformative breakthroughs, faces notable technical challenges. One major hurdle is achieving precise control at the molecular scale. Current limitations involve the difficulty in manipulating individual molecules due to their size and intricacies. However, emerging solutions, such as advanced algorithms and artificial intelligence, show promise in overcoming these challenges. For instance, deep reinforcement learning (DRL) has been applied to enhance manipulation capabilities in micro-electro-mechanical systems (MEMS) and nanotechnology, showcasing the potential of smart algorithms in addressing technical limitations. Another technical challenge is the potential toxicity of nanomaterials. Understanding and mitigating the adverse effects on biological systems are essential. Research, as seen in case studies, explores the applications and toxicity of nanoparticles in real-world scenarios, providing insights into the balance between innovation and safety.

Ethical Implications: Responsible Use and Future Considerations

The ethical considerations in molecular nanotechnology are paramount. Responsible use of molecular manipulation raises questions about the unbiased determination of hazards and risks. Ethical principles like nonmaleficence, autonomy, justice, and privacy become crucial in ensuring the safe and equitable development of nanotechnology. As molecular manipulation capabilities advance, ethical discussions extend to the "god-like" powers conferred by nanotechnology. The potential to eliminate ethical issues, such as assembling synthetic meat, highlights the need for careful consideration of the societal, environmental, and ethical impacts. Public trust and the environmental impact of nanotechnologies also pose ethical challenges.

Nanotechnology's implications necessitate ongoing dialogue and transparent communication to build trust and address concerns. The potential risks and benefits must be carefully weighed, fostering responsible innovation that aligns with societal values. Addressing the technical challenges in molecular nanotechnology requires innovative solutions, leveraging advancements in algorithms and materials science. Simultaneously, ethical considerations demand responsible development and transparent communication to navigate the complex landscape of nanotechnology's societal impacts.

Fundamentals of Matter Manipulation in Nanoscience

The chapter on "Fundamentals of Matter Manipulation" delves into the intricate world of nanoscience, exploring the design, fabrication, and manipulation of structures and devices with atomic precision. As we recapitulate key concepts, it becomes evident that current nanoscience developments are marked by significant strides in understanding and harnessing matter at the atomic and molecular scales.

Current Nanoscience Developments
1. Convergence of Sciences:

Nanoscience emerges as a convergence of physics, materials science, and biology, fostering a multidisciplinary approach to manipulate materials at the atomic and molecular scales.

2. Technological Tools:

Tools like scanning tunnelling microscopes (STMs) and atomic force microscopes (AFMs) play a pivotal role in manipulating matter, allowing researchers to visualize and control structures at unprecedented scales.

3. Structural DNA Nanotechnology:

Innovations like structural DNA nanotechnology showcase the state of the art, introducing shape-changing structural modules for precise manipulation at the nanometer scale.

Future Developments in Matter Manipulation

As we look ahead, the future of matter manipulation in nanoscience holds great promise.

1. Advanced Algorithms:

The integration of advanced algorithms and artificial intelligence is anticipated to revolutionize molecular manipulation. Deep reinforcement learning (DRL) is already enhancing manipulation capabilities in micro-electro-mechanical systems (MEMS) and nanotechnology.

2. Nanotechnology and Regenerative Medicine:

Future developments may address challenges in overcoming barriers to the brain and eye, presenting new possibilities for regenerative medicine through nanotechnology.

3. Biotechnological Applications:

The intersection of nanotechnology with biotechnology is likely to lead to novel applications, further expanding the realm of possibilities in molecular manipulation.

In this chapter we therefore underscore the dynamic landscape of nanoscience. Current developments showcase our growing mastery over manipulating matter at the nanoscale, while the future promises even more sophisticated techniques and applications. As nanoscience continues to evolve, it opens avenues for groundbreaking innovations that will undoubtedly reshape our understanding and utilization of matter in the years to come.

4. Nanomaterials: Building Blocks of the Future

Introduction to Nanomaterials

Nanomaterials, by definition, possess at least one external dimension measuring 1-100 nanometres (nm). This scale places them in the nanoscale range, where unique properties and behaviours emerge due to quantum effects and increased surface area-to-volume ratios. The significance of nanomaterials reverberates across various industries, influencing technological advancements, materials science, and healthcare.

Defining Nanomaterials

In ISO/TS 80004, nanomaterial is characterized as a material with any external dimension in the nanoscale or having internal and surface structures in the nanoscale range. This broad definition encompasses a diverse array of materials, including nanoparticles, nanocomposites, and nanostructured materials.

Scope of Nanomaterials

The scope of nanomaterials extends beyond traditional materials, offering unprecedented possibilities in multiple industries:

1. Materials Science:

Nanomaterials exhibit unique mechanical, electrical, and optical properties that deviate from their bulk counterparts. These distinctive characteristics open avenues for designing advanced materials with enhanced performance and novel functionalities.

2. *Electronics and Technology:*

In the electronics industry, nanomaterials facilitate the development of smaller, more efficient components. Quantum dots, for example, are nanoscale semiconductor particles with applications in high-resolution displays and solar cells.

3. *Medicine and Healthcare:*

Nanomaterials play a pivotal role in drug delivery, diagnostics, and imaging. Nanoparticles can be engineered to deliver therapeutic agents precisely to target cells, minimizing side effects. Additionally, nanomaterials enable the development of advanced imaging techniques for early disease detection.

4. *Energy:*

Nanomaterials contribute to the evolution of energy storage and conversion devices. Nanocomposites enhance the efficiency of batteries and supercapacitors, while nanomaterials like graphene hold promise for improved solar cells and fuel cells.

5. *Environmental Applications:*

Nanomaterials find applications in environmental remediation and monitoring. Nanoparticles can be used to remove pollutants from water, and nanosensors enable real-time monitoring of environmental parameters.

6. Manufacturing and Construction:

Nanomaterials enhance the mechanical properties of materials used in manufacturing and construction. Nano-reinforcements in composites lead to stronger and lighter materials, contributing to the development of advanced structural components.

As we navigate the nanoscale world, nanomaterials emerge as versatile building blocks with immense potential. The interdisciplinary nature of nanomaterial research fosters collaboration among scientists, engineers, and professionals, driving innovation across diverse industries. This introduction sets the stage to explore the myriad applications and challenges associated with harnessing the power of nanomaterials in our ever-evolving technological landscape.

Types of Nanomaterials
1. Nanoparticles
Synthesis Methods:

Nanoparticles, defined by their dimensions ranging from 1 to 100 nanometers, exhibit unique properties due to their quantum effects and increased surface area-to-volume ratios. Various synthesis methods, such as chemical reduction, sol-gel processes, and green synthesis using biological agents, enable the controlled production of nanoparticles.

Properties:

Nanoparticles showcase distinctive properties, including size-dependent optical, magnetic and catalytic behaviours. Their large surface area enhances reactivity, making them ideal for applications in drug delivery and imaging.

Applications:

In drug delivery, nanoparticles enable targeted therapy, delivering therapeutic agents precisely to affected cells while minimizing side effects. Additionally, in imaging, nanoparticles serve as contrast agents, enhancing the visibility of specific structures in diagnostic techniques.

2. Nanocomposites
Blending at the Nanoscale:

Nanocomposites result from blending different materials at the nanoscale, creating hybrid structures with enhanced mechanical, thermal, and electrical properties. Common techniques involve incorporating nanoparticles into polymers, ceramics, or metals.

Enhanced Properties:

The synergy between materials at the nanoscale imparts nanocomposites with improved mechanical strength, thermal stability, and electrical conductivity compared to their bulk counterparts. These enhancements find applications in aerospace, automotive, and construction industries.

3. Nanowires and Nanotubes
Structural Characteristics:

Nanowires and nanotubes are one-dimensional nanostructures with diameters in the nanometer range. Nanowires are solid, elongated structures, while nanotubes possess hollow cylindrical shapes. Carbon nanotubes, for example, exhibit exceptional structural strength.

Applications and Role in Electronics:

Nanowires and nanotubes play a crucial role in electronics and nanodevices. Their unique electronic properties make them suitable for components in nanoscale electronics. Nanotubes, in particular, are explored for applications in lightweight and flexible electronics, serving as conductive elements in various devices.

The fascinating realm of nanomaterials encompasses diverse types, each with its distinct characteristics and applications. From the precision of nanoparticles in healthcare to the enhanced properties of nanocomposites and the unique electronic roles of nanowires and nanotubes, nanomaterials continue to revolutionize various industries, promising advancements in technology, medicine, and materials science.

Fabrication and Characterization of Nanomaterials
Synthesis Techniques
1. Bottom-Up and Top-Down Approaches

Nanomaterials can be synthesized through two main approaches. Bottom-up methods involve building materials atom by atom or molecule by molecule to create nanostructures. In contrast, top-down methods start with bulk materials, and through various techniques, reduce them to nanoscale dimensions.

2. Chemical Vapour Deposition (CVD)

CVD is a prevalent technique where vaporized precursor materials react on a substrate, forming a thin film. This method is widely used for the synthesis of nanomaterials like graphene and carbon nanotubes, offering control over the structure and properties.

3. Sol-Gel Method

The sol-gel process involves the conversion of a colloidal solution (sol) into a gel, followed by drying and heat treatment to form the final material. This technique is versatile, allowing the synthesis of various nanomaterials, particularly metal oxides and hybrid organic-inorganic materials.

4. Other Methods

Numerous additional methods, including ball milling, hydrothermal synthesis, and electrospinning, contribute to the diverse array of nanomaterial fabrication techniques. Each method offers specific advantages depending on the desired material and application.

Analytical Techniques

1. Scanning Electron Microscopy (SEM)

SEM provides high-resolution images of nanomaterial surfaces, offering insights into their morphology. This technique is crucial for visualizing the size, shape, and distribution of nanoparticles.

2. X-ray Diffraction (XRD)

XRD is employed to determine the crystal structure of nanomaterials. By analysing X-ray diffraction patterns, researchers can identify the type of crystal lattice and gain information about the material's composition.

3. Spectroscopy

Various spectroscopic techniques, such as UV-Vis, IR, and Raman spectroscopy, enable the study of nanomaterials' electronic and vibrational properties. These techniques provide information about chemical composition, functional groups, and bonding states.

Understanding Nanomaterial Structure and Composition

The combination of synthesis and analytical techniques empowers researchers to understand nanomaterial structure and composition comprehensively. By controlling fabrication processes, scientists can tailor nanomaterial properties for specific applications. Analytical techniques, on the other hand, offer a detailed look into the resulting structures, allowing for precise characterization and quality assessment.

In summary, the synergy between fabrication and characterization techniques is fundamental to advancing our understanding and utilization of nanomaterials, paving the way for innovative applications across diverse industries.

Properties of Nanomaterials

Nanomaterials exhibit a rich tapestry of properties that distinguish them from bulk materials, unlocking a wide range of applications in various fields.

Size-Dependent Properties
1. Quantum Effects:

As nanomaterials approach the nanoscale, quantum effects become pronounced. Quantum confinement leads to unique electronic properties, affecting conductivity, band structure, and optical behaviour. This phenomenon is crucial in applications like quantum dots for imaging and quantum computing.

2. Surface Area:

The high surface area-to-volume ratio of nanomaterials enhances their reactivity and interaction with surrounding environments. This property is exploited in catalysis, drug delivery, and sensors, where increased surface area improves efficiency.

Mechanical and Thermal Characteristics
1. Strength:

Nanomaterials often exhibit exceptional strength due to their unique structures. This strength finds applications in materials science, enabling the development of lightweight yet strong materials for various industries.

2. Flexibility:

Nanostructures can display remarkable flexibility, allowing for tailored materials that bend and flex without losing structural integrity. This property is valuable in applications like flexible electronics and wearable technologies.

3. Thermal Conductivity:

Nanomaterials, with their reduced dimensions, can possess enhanced thermal conductivity. This property is advantageous in fields such as thermal management, where nanomaterials contribute to efficient heat dissipation.

Optical and Electronic Properties
1. Optical Properties:

The optical properties of nanomaterials are heavily influenced by their size and structure. This leads to tuneable colours, light absorption, and emission properties. Applications include sensors, imaging, and light-emitting devices.

2. *Electronic Properties:*

Size-dependent electronic properties enable the development of nanoelectronic devices. Nanomaterials are crucial in the semiconductor industry, facilitating the creation of smaller, faster, and more efficient electronic components.

In summary, nanomaterials' properties are intricately linked to their size, rendering them versatile for a myriad of applications. Understanding and harnessing these properties open doors to innovations in medicine, electronics, energy, and beyond.

Diverse Applications of Nanotechnology

Nanotechnology, with its ability to manipulate matter at the nanoscale, has found diverse applications across various industries, showcasing its versatility and potential impact on medicine, electronics, and the environment.

Medicine and Healthcare
1. *Drug Delivery Systems:*

Nanoparticles serve as efficient carriers for drug delivery. Their small size facilitates targeted delivery, reducing side effects and improving drug efficacy. This innovation has revolutionized treatments for various diseases, providing more effective and personalized therapeutic options.

2. Imaging Agents:

Nanomaterials act as contrast agents in medical imaging, enhancing the visibility of tissues and organs. This application is crucial for accurate diagnosis and monitoring of diseases. Nanoparticles, such as quantum dots, offer superior imaging capabilities, contributing to advancements in medical diagnostics.

Electronics and Nanodevices
1. Transistors and Memory Devices:

Nanotechnology has enabled the development of nanoscale transistors and memory devices. The miniaturization of electronic components enhances device performance, leading to faster and more efficient electronics. This progress is fundamental to the continuous evolution of electronic devices.

2. Nanosensors

Nanosensors utilize the unique properties of nanomaterials for highly sensitive and selective detection of various substances. These sensors find applications in healthcare for diagnostics, environmental monitoring, and even in everyday consumer products, enhancing safety and efficiency.

Energy and Environment
1. Solar Cells

Nanotechnology plays a pivotal role in improving the efficiency of solar cells. Nanomaterials, such as quantum dots and nanowires, enhance light absorption and electron transport, contributing to the development of high-performance solar energy harvesting systems.

2. Catalysis

Nanocatalysts exhibit enhanced catalytic activity due to their high surface area and unique structures. This property is harnessed in catalysis for efficient chemical reactions with reduced energy consumption. Nanocatalysts contribute to the development of sustainable and green processes.

3. Water Purification:

Nanotechnology contributes to the creation of advanced water purification technologies. Nanomaterials like nanoparticles and nanocomposites are effective in removing contaminants, providing solutions for clean and safe water supplies. This application addresses critical environmental and public health challenges.

In summary, the diverse applications of nanotechnology underscore its transformative potential across critical sectors. From advancing healthcare to revolutionizing electronics and addressing environmental concerns, nanotechnology continues to pave the way for innovative solutions.

Challenges and Future Perspectives in Nanoscience

Nanoscience, while offering tremendous advancements, faces challenges concerning safety, ethics, and environmental impact. Simultaneously, ongoing research points towards exciting future trends and innovations.

Safety and Ethical Concerns
1. Nanotoxicity:

Nanoparticles, due to their unique properties, may pose potential health risks. Nanotoxicity studies aim to understand and mitigate adverse effects on human health and the environment. Researchers are exploring the safe use of nanomaterials, considering their potential impacts on living organisms.

2. *Environmental Impact:*

The widespread application of nanomaterials raises concerns about their environmental repercussions. Studies delve into understanding the fate and transport of nanoparticles, their accumulation in ecosystems, and potential long-term effects on biodiversity. Balancing technological progress with environmental stewardship is crucial for sustainable nanoscience development.

Future Trends and Innovations
1. Emerging Applications:

Nanoscience continues to drive innovation across diverse fields. In medicine, nanotechnology promises targeted drug delivery, personalized treatments, and advanced diagnostics. In electronics, the development of nanoscale transistors and sensors opens new possibilities for faster and more efficient devices. Energy applications, such as advanced solar cells and nanocatalysts for clean processes, showcase the potential for sustainable solutions.

1. 2. Research Directions:

Ongoing research focuses on harnessing nanoscience for societal benefit. Biocompatible nanomaterials, designed for medical applications, aim to revolutionize treatments with minimal side effects. Nanosensors for environmental monitoring provide real-time data for better decision-making. Nanocatalysis research explores efficient and eco-friendly processes for industries.

3. Ethical Considerations:

As nanoscience progresses, ethical considerations become paramount. Transparent communication, responsible innovation, and inclusivity in decision-making processes are vital. Addressing ethical concerns ensures the responsible development and deployment of nanotechnology.

In summary, the challenges and future perspectives in nanoscience are intertwined. Striking a balance between innovation and safety, and addressing ethical considerations, will shape the trajectory of nanoscience toward a sustainable and impactful future.

Nanomaterial Applications and Lessons Learned
Success Stories: Real-World Impact
1. Sunscreen Enhancement:

Nanotechnology revolutionized the cosmetics industry by incorporating nanoparticles into sunscreens. These tiny particles provide better UV protection, enhancing the effectiveness of sunscreens and reducing the need for thicker applications. This implementation has significantly improved skin protection and has become a common feature in various skincare products.

2. Biomedical Breakthroughs:

Nanotechnology has played a pivotal role in biomedical applications. Engineered nanomaterials have shown success in drug delivery systems, targeting specific cells and minimizing side effects. This has transformed cancer treatment, enabling more effective therapies with reduced harm to healthy tissues. The impact on healthcare is profound, showcasing the potential of nanomaterials in improving treatment outcomes.

3. Energy Storage Advancements:

Recent studies have demonstrated the successful use of nanomaterials in overcoming challenges in energy storage. Specifically, the application of nanomaterials in microbial electrochemical cells (MECs) has addressed issues such as high material costs and low efficiency. The use of nanomaterials has enhanced the performance of MECs, paving the way for more sustainable and cost-effective energy solutions.

Lessons Learned: Overcoming Challenges
1. Translation to Clinical Practice:

The field of nanomedicine faced challenges in translating laboratory innovations into commercially viable medical products. Grand challenges in nanomedicine highlighted the need for addressing safety concerns and ensuring the successful transition from the laboratory to clinical applications. Researchers have learned valuable lessons about navigating regulatory landscapes and optimizing processes for safe and effective clinical translation.

2. Environmental Impact Considerations:

As nanotechnology progresses, lessons have been learned about the potential environmental impacts of nanomaterials. Early warnings and assessments have provided insights into managing risks associated with nanoparticle release. These lessons emphasize the importance of responsible development, environmental monitoring, and sustainable practices to minimize adverse effects.

In summary, the success stories of nanomaterial applications in sunscreen enhancement, biomedical breakthroughs, and energy storage advancements highlight the transformative impact of nanotechnology. Lessons learned from challenges in clinical translation and environmental impact considerations underscore the importance of responsible innovation and ongoing research to maximize the benefits of nanomaterial applications.

Nanomaterials - Building Blocks of the Future

In concluding the chapter "Nanomaterials: Building Blocks of the Future," we recapitulate key concepts that underscore the transformative potential of nanomaterials. This chapter has provided a comprehensive overview of different classes of nanomaterials, covering their synthesis, fabrication, characterization, and unique properties.

Recapitulation of Key Concepts:
1. *Diverse Nanomaterial Classes:*

The exploration encompassed various nanomaterial classes, each with distinct properties and applications. This diversity includes nanotubes, nanomagnets, nanowires, and carbon nanoparticles, illustrating the broad spectrum of nanomaterials that form the foundation of nanoscience and technology.

2. *Synthesis and Functionalization:*

The synthesis and functionalization processes were addressed, shedding light on the methods employed to create and tailor nanomaterials for specific applications. Understanding these processes is crucial for harnessing the full potential of nanomaterials in diverse fields.

3. Versatile Applications:

Nanomaterials have found applications in various domains, including medicine, energy, and electronics. Their unique properties enable advancements such as drug delivery systems, energy storage solutions, and novel electronic devices.

Anticipated Role of Nanomaterials in Shaping the Future:

As we peer into the future, nanomaterials are poised to play a pivotal role in shaping technological landscapes and societal progress. The anticipated contributions include:

1. Revolutionizing Medicine:

Nanomaterials hold the promise of revolutionizing medicine through targeted drug delivery and enhanced diagnostic techniques. Their ability to interact at the nanoscale allows for unprecedented precision in medical interventions, potentially transforming treatment approaches and improving patient outcomes.

2. Advancing Sustainable Technologies:

Nanomaterials are key players in advancing sustainable technologies. From energy storage solutions to environmental remediation, their unique properties contribute to the development of eco-friendly technologies, addressing global challenges related to climate change and resource management.

3. Innovations in Electronics and Beyond:

The use of nanomaterials in electronics is expected to lead to breakthroughs in device miniaturization, efficiency, and performance. Beyond electronics, nanomaterials will likely contribute to innovations in diverse fields, from materials science to catalysis, opening new frontiers in research and development.

In summary, the chapter has laid the foundation for understanding the vast potential of nanomaterials. Their versatile applications and anticipated roles in medicine, sustainability, and innovation position nanomaterials as true building blocks of the future.

5. Molecular Nanotechnology: A Revolution Unveiled

Introduction

Molecular Nanotechnology (MNT) emerges as a groundbreaking frontier, promising a revolution in our approach to manipulating matter at the molecular and atomic levels through mechanosynthesis. This chapter embarks on an exploration of MNT, unravelling its definition, evolution, and profound significance across various domains.

Definition of Molecular Nanotechnology

Molecular Nanotechnology is a transformative technology that leverages mechanosynthesis to construct structures with atomic precision. This entails the ability to design and build intricate materials, devices and systems at the molecular scale, allowing unprecedented control over the properties and functionalities of matter. The foundation of MNT lies in the precise manipulation of individual molecules, enabling the creation of materials with tailored characteristics for diverse applications.

Evolution of Molecular Nanotechnology

The evolution of Molecular Nanotechnology is intrinsically tied to the broader development of nanotechnology. Initially conceptualized as a theoretical framework, nanotechnology gained momentum with scientific and engineering advancements, leading to practical applications in various fields. Molecular

Nanotechnology emerged as a specialized subset, driven by the ambition to manipulate matter at the atomic and molecular levels. The journey from theoretical concepts to tangible breakthroughs reflects the evolution of nanotechnology, with a focus on achieving molecular precision. As the field advanced, the term "nanotechnology" evolved to describe the manipulation of matter at the nanoscale, encompassing chemical, molecular, and supramolecular manipulation.

Significance of Molecular Nanotechnology

The significance of Molecular Nanotechnology spans across diverse domains, reshaping industries and opening new frontiers in scientific exploration. Its precision at the molecular level allows for the creation of advanced materials with tailored properties, revolutionizing fields such as medicine, electronics and materials science. The applications of nanotechnology have played a pivotal role in the modernization of various industries and its potential extends well beyond industrial applications. It holds promise in fields like medicine, where precise drug delivery systems and targeted therapies can be developed with unparalleled accuracy. This level of control over material properties and functions unlocks innovative solutions to societal challenges, making Molecular Nanotechnology a catalyst for progress.

So, as we embark on our journey unfolding the intricacies of Molecular Nanotechnology, its significance will become increasingly evident. This exploration promises to unveil the full potential of MNT, propelling us into an era where precision at the molecular level becomes the cornerstone of innovation.

Foundation of Molecular Nanotechnology

Molecular Nanotechnology (MNT) is rooted in the ability to manipulate matter at the nanoscale, allowing for precision at the molecular level. Nanoscale manipulation involves techniques that enable the observation, movement, and manipulation of materials at atomic and molecular scales. These techniques form the foundation of MNT, providing the means to engineer functional structures atom by atom.

Nanoscale Manipulation Techniques
1. Scanning Probe Microscopy (SPM):

SPM techniques, including Atomic Force Microscopy (AFM) and Scanning Tunnelling Microscopy (STM), enable researchers to visualize and manipulate individual atoms. AFM uses a sharp tip to sense forces at the atomic level, while STM relies on the quantum tunnelling effect to measure the distance between the tip and the sample surface.

2. Nanomanipulation via Laser Trapping:

Laser trapping involves the use of focused laser beams in order to trap and manipulate nanoscale objects. This technique allows for the precise control of particles, facilitating assembly at the molecular level.

3. Chemical Vapour Deposition (CVD):

CVD is a technique used for thin-film deposition at the atomic scale. It involves the introduction of precursor gases, which react on a substrate surface to form a thin, precisely controlled layer of material. This technique is crucial for building nanoscale structures with atomic precision.

Molecular Assembly: Building Atom by Atom

1. DNA Origami:

Inspired by the self-assembling properties of DNA, scientists leverage DNA strands as templates for assembling nanoscale structures. Through careful design, DNA origami allows the creation of intricate shapes and functional nanostructures.

2. Molecular Beam Epitaxy (MBE):

MBE is a method for depositing thin films with atomic precision. It involves the deposition of atoms or molecules onto a substrate, allowing for the controlled growth of crystalline structures. MBE is vital for fabricating semiconductors and other nanoscale electronic components.

3. Supramolecular Chemistry:

Supramolecular chemistry focuses on non-covalent interactions between molecules to assemble larger, more complex structures. This approach enables the creation of functional materials through the precise arrangement of molecular building blocks.

The foundation of molecular nanotechnology, characterized by nanoscale manipulation and molecular assembly techniques, empowers scientists and engineers to construct functional structures atom by atom. As we delve deeper into these methods, the promise of Molecular Nanotechnology unfolds, offering unparalleled control over matter and the potential for revolutionary advancements in various different fields.

Transformative Impact on Industries

Molecular Nanotechnology (MNT) is revolutionizing various industries, bringing about transformative changes in medicine and healthcare, electronics and computing, as well as materials and manufacturing.

Medicine and Healthcare
1. *Drug Delivery Systems:*

MNT enables the design and creation of highly efficient drug delivery systems. Nano-sized carriers can transport drugs precisely to targeted cells, improving therapeutic efficacy while minimizing side effects.

2. *Targeted Therapies:*

Molecular precision allows for the development of targeted therapies, tailoring treatments to individual patients based on their genetic makeup. This approach enhances treatment outcomes and reduces adverse effects.

3. *Nanorobots for Medical Procedures:*

MNT introduces the concept of nanorobots capable of performing medical procedures at the molecular level. These miniature devices hold the potential for precise surgeries and targeted interventions.

Electronics and Computing
1. *Ultra-Miniaturized Components:*

MNT allows for the creation of ultra-miniaturized electronic components, facilitating the development of smaller and more powerful devices. This miniaturization enhances the efficiency and performance of electronic systems.

2. Quantum Computing:

The precision offered by MNT is instrumental in the development of quantum computing. Quantum bits (qubits) can be manipulated at the molecular level, enabling quantum computing's unparalleled processing power.

3. Molecular Electronics and Nanoscale Circuits:

MNT plays a crucial role in the advancement of molecular electronics, paving the way for nanoscale circuits. This leads to the development of faster and more energy-efficient electronic devices.

Materials and Manufacturing
1. Stronger and Lightweight Materials:

MNT contributes to the creation of materials with exceptional strength and lightweight properties. Nanoscale structures allow for enhanced material properties, benefiting industries such as aerospace and construction.

2. Nanoscale 3D Printing and Manufacturing:

MNT facilitates nanoscale 3D printing and manufacturing processes, enabling the production of intricate structures with unprecedented precision. This advancement has implications for creating customized products, from medical implants to electronic components.

The transformative impact of Molecular Nanotechnology extends across all these industries, ushering in an era of unprecedented innovation and efficiency.

Molecular Nanotechnology's Pioneering Role in Water Purification and Energy Harvesting

Molecular Nanotechnology (MNT) stands at the forefront of environmental innovation, wielding transformative capabilities in water purification and energy harvesting. These applications showcase MNT's potential to address critical challenges and propel us towards a sustainable future.

Water Purification: Nanotech-Enabled Filtration and Desalination

1. Nanotech-Enabled Filtration:

MNT introduces nanoscale filtration methods that revolutionize water treatment. By leveraging the adsorption properties of nanomaterials, contaminants are efficiently removed, enhancing the purity of water sources.

2. Desalination Breakthroughs:

Molecular Nanotechnology plays a pivotal role in advancing desalination techniques. Tailored nanomaterials with selective permeability allow for efficient desalination processes, addressing freshwater scarcity by converting seawater into a viable resource.

Energy Harvesting: Nanomaterials and Nanogenerators
1. Efficient Solar Cells:

MNT contributes to the development of nanomaterials that redefine the efficiency of solar cells. These nanomaterials enhance light absorption, optimizing the conversion of sunlight into electricity and promoting the widespread adoption of sustainable solar energy.

2. Nanogenerators for Sustainable Power:

Molecular Nanotechnology pioneers the creation of nanogenerators, innovative devices that harvest energy from the environment. These nanogenerators tap into mechanical vibrations at the nanoscale, providing a novel and sustainable source of power for various applications.

In summary, Molecular Nanotechnology emerges as a beacon of hope for environmental challenges. Through nanotech-enabled filtration, desalination breakthroughs and advancements in energy harvesting technologies, MNT offers a revolutionary toolkit to safeguard our planet. As we continue to explore and harness the potential of molecular precision, the environmental landscape is destined for positive transformation, paving the way for a cleaner and more sustainable world.

Societal Implications and Ethical Considerations of Molecular Nanotechnology

Molecular Nanotechnology (MNT) holds immense promise, but its transformative power necessitates a vigilant exploration of societal implications and ethical considerations. As we delve into the nanoscale, a careful balance between innovation and ethical responsibility becomes imperative.

Potential Risks: Environmental and Health Concerns

1. Environmental Impact:

The manufacturing processes involved in MNT could lead to the release of nanomaterials into the environment. The long-term effects of these novel materials on ecosystems raise concerns about biodiversity and ecological balance.

2. Health Concerns:

The introduction of nanoscale materials raises questions about their impact on human health. The potential for unintended consequences, such as toxicity or unknown biological interactions, requires a thorough examination in order to ensure the safety of both consumers and workers.

Ethical Guidelines: Responsible Development and Deployment

1. Unbiased Risk Assessment:

Ethical considerations involve the unbiased determination of hazards and risks associated with MNT. Rigorous risk assessments should precede the development and deployment of nanotechnological applications, ensuring the well-informed decision-making process.

2. Nonmaleficence and Autonomy:

A commitment to "do no harm" (nonmaleficence) and respect for individual autonomy are ethical imperatives. Striking a balance between scientific progress and ethical values ensures that MNT advancements prioritize human well-being and individual freedoms.

3. Justice and Fairness:

Fairness within nanotechnology industries is crucial. Investments in appropriate safeguards, transparent governance, and equitable access to benefits and risks must be prioritized in order to prevent the exacerbation of existing social disparities.

4. Privacy Considerations:

As MNT introduces new technologies and applications, privacy concerns may arise. Ethical guidelines should address the responsible handling of personal information, preventing unauthorized use or unintended consequences related to privacy breaches.

In summary, the extraordinary capabilities of Molecular Nanotechnology come with the responsibility to navigate ethical landscapes diligently. By addressing potential risks, conducting unbiased assessments, and adhering to ethical guidelines, we pave the way for a future where MNT contributes to society while upholding the highest standards of ethical conduct.

Illuminating the Nano Frontiers: Global Innovations and Collaborations in Molecular Nanotechnology

Molecular Nanotechnology (MNT) has become a focal point of global innovation, with various countries at the forefront, leading the charge in groundbreaking advancements. As we traverse the nanoscale landscape, these pioneering nations are not only propelling their scientific communities forward but also engaging in international collaborations to address global challenges collectively.

Pioneering Countries in Molecular Nanotechnology:
1. United States:

With robust investments in research and development, the United States stands as a pioneer in molecular nanotechnology. Leading academic institutions and private enterprises contribute significantly to advancements in nanoscale science and engineering.

2. China:

Rapidly emerging as a global leader, China's dedication to nanotechnology research and application is evident in its sustained investment and comprehensive national strategies. Collaborations between academia, industry, and government institutions drive China's prominence in the nanotech landscape.

3. Germany:

Renowned for its emphasis on applied research, Germany is a key player in molecular nanotechnology. Collaborations between German research institutions and industry partners foster innovation, contributing to the country's prominent position in nanoscale science.

International Collaborations Addressing Global Challenges:
1. Global Collaborations in Research:

The collaborative nature of nanotechnology is evident through international research initiatives. Countries with diverse expertise pool their resources and knowledge to tackle complex challenges, ranging from environmental sustainability to healthcare.

2. Multinational Projects and Alliances:

Initiatives like the European Union's Horizon 2020 program foster collaborations across borders. Such alliances bring together researchers, engineers, and policymakers to address societal challenges and leverage the transformative potential of molecular nanotechnology.

3. Knowledge Exchange and Networking:

Global conferences and symposiums serve as platforms for scientists and innovators to exchange knowledge. These events facilitate networking, encouraging cross-border collaborations that transcend geographical boundaries and contribute to the collective progress of molecular nanotechnology.

The fusion of national efforts and international collaborations in molecular nanotechnology propels the field forward, driving innovation, and addressing global challenges. As nations unite in their pursuit of nanoscale breakthroughs, the future promises an era of unprecedented collaboration, pushing the boundaries of what is possible at the molecular level.

Future Prospects and Challenges in Molecular Nanotechnology

Molecular Nanotechnology (MNT) stands at the cusp of revolutionizing diverse sectors, offering unparalleled prospects and posing intriguing challenges. The future promises groundbreaking developments, with researchers exploring emerging trends and breakthroughs made in order to enhance the potential of MNT.

Emerging Trends and Anticipated Breakthroughs:

1. Biomedical Applications:

Anticipated breakthroughs in MNT include novel drug delivery systems, precise targeting of cancer cells and advancements in personalized medicine. Researchers are exploring nanoscale interventions for disease treatment, enabling more effective and tailored therapeutic approaches.

2. Quantum Computing with Nanoscale Components:

The integration of molecular nanotechnology in quantum computing is a burgeoning trend. Scientists are working on utilizing nanoscale components for quantum bits (qubits), paving the way for faster and more efficient quantum information processing.

3. Environmental Remediation:

MNT holds the potential for addressing environmental challenges, including efficient water purification and pollution control. Nanoscale materials can be engineered for targeted removal of contaminants, offering sustainable solutions for environmental conservation.

Overcoming Challenges:
1. Safety and Toxicity Concerns:

A primary challenge involves ensuring the safety of nanomaterials. Researchers are employing advanced testing methods and computational models in order to assess the toxicity of nanoparticles, mitigating potential health and environmental risks.

2. Standardization and Regulation:

The diverse applications of MNT necessitate robust standardization and regulatory frameworks. Ongoing research focuses on establishing industry standards and guidelines in order to ensure responsible development and deployment of nanotechnological innovations.

3. Ethical Considerations:

As MNT advances, ethical considerations become paramount. Researchers are actively engaging in discussions around ethical frameworks used in order to guide the responsible use of nanotechnology, addressing concerns related to privacy, autonomy, and societal impacts.

Diverse Research Avenues to Enhance Molecular Nanotechnology:
1. Materials Science and Engineering:

Advances in materials science play a pivotal role in enhancing MNT. Researchers are exploring innovative nanomaterials with tailored properties, allowing for precise control over molecular structures and functionalities.

2. Interdisciplinary Collaboration:

Collaborations between nanotechnologists, biologists, physicists, and engineers are fostering interdisciplinary approaches. These collaborations enable a holistic understanding of nanoscale phenomena and drive the development of multifaceted applications.

3. Computational Modelling:

Computational tools are pivotal in designing and optimizing nanoscale structures. Researchers employ sophisticated modelling techniques in order to predict the behaviour of nanomaterials, facilitating the rational design of novel molecular structures.

In the dynamic landscape of molecular nanotechnology, the future holds immense promise. By addressing challenges through rigorous research, ethical considerations, and collaborative efforts, scientists are poised to unlock unprecedented possibilities, shaping a future where molecular precision transforms the way we approach medicine, technology, and environmental sustainability.

A Glance into Molecular Nanotechnology's Triumphs and Tomorrow

Molecular Nanotechnology (MNT) has evolved from a visionary concept to a transformative force, rewriting the possibilities in science and industry. As we stand at the precipice of the nano frontier, it's essential to recapitulate the key insights that underscore its ongoing impact and immense future potential.

Ongoing Impact:
1. Revolutionizing Industries:

Contrary to its miniature name, MNT has unleashed macroscopic changes, revolutionizing industries globally. From healthcare to manufacturing, the precision at the molecular level has led to unparalleled advancements.

2. Innovative Materials:

MNT has birthed a new era of materials science. Innovative nanomaterials with tailored properties and functionalities have become the building blocks for everything from lightweight and durable materials to advanced electronics.

3. Biomedical Breakthroughs:

The impact of MNT in biomedical applications cannot be overstated. Precision drug delivery, targeted cancer treatments, and advanced diagnostic tools are just glimpses of its contributions to the field of medicine.

Future Potential:
1. Quantum Computing Revolution:

As we look forward, MNT is poised to play a pivotal role in the quantum computing revolution. The integration of nanoscale components in quantum systems holds the promise of computing power previously thought to be science fiction.

2. Environmental Stewardship:

MNT's future potential extends to environmental stewardship. Nanotechnology offers solutions for efficient water purification, pollution control, and sustainable energy, addressing critical challenges for a healthier planet.

3. *Customized Medicine:*

The dawn of personalized medicine beckons, with MNT driving the charge. Tailoring treatments to an individual's molecular profile opens avenues for more effective, targeted, and less invasive medical interventions.

Acknowledging Challenges:

1. *Ethical Considerations:*

The journey of MNT is not without its ethical considerations. As it progresses, ethical frameworks must evolve to ensure responsible development, addressing concerns related to privacy, autonomy, and societal impact.

2. *Safety Assurance:*

Ensuring the safety of nanomaterials remains a paramount challenge. Rigorous testing and regulatory frameworks are imperative to mitigate potential health and environmental risks associated with nanotechnology.

Molecular Nanotechnology has transcended the realms of imagination, leaving an indelible mark on the scientific landscape. Its ongoing impact reshapes industries, and its future potential unveils unprecedented opportunities. As we navigate the challenges, the nano frontier promises a future where precision at the molecular level transforms not just industries but the very fabric of our existence.

6. Nanoelectronics and Computing

Navigating the Quantum Realm: A Historical Voyage through Nanoelectronics and Computing

Welcome to a chapter that unravels the intricate tapestry of nanoelectronics and computing, where the minuscule meets the monumental. Nanoelectronics, at its core, is the study and application of electronic components and devices at the nanoscale, pushing the boundaries of traditional electronics to the quantum realm.

Defining Nanoelectronics:

Nanoelectronics represents a paradigm shift from conventional electronics by harnessing the unique properties of materials at the nanoscale. This field delves into the design, fabrication and application of electronic circuits and components on the order of nanometres, allowing for unprecedented control and efficiency in information processing.

Historical Evolution:

The journey of nanoelectronics traces back to the latter half of the 20th century, marked by groundbreaking discoveries and milestones that paved the way for today's cutting-edge technologies.

1. Birth of Transistors:

The transistor, a cornerstone of electronic devices, saw a revolution in the late 1940s. As transistors shrank to smaller dimensions, researchers glimpsed the potential of manipulating individual atoms for computing purposes.

2. Moore's Law:

Coined by Gordon Moore in 1965, Moore's Law predicted that the number of transistors on a microchip would double approximately every two years, driving an exponential increase in computing power. This law became a guiding principle for the semiconductor industry.

3. Invention of Scanning Tunnelling Microscope (STM):

The 1980s witnessed the invention of the STM, a groundbreaking tool allowing scientists to visualize and manipulate individual atoms. This marked a significant leap towards the precision required for nanoelectronics.

4. Emergence of Nanotechnology:

The late 20th century witnessed the rise of nanotechnology, a multidisciplinary field that intersects with nanoelectronics. Pioneering research explored the fabrication of devices at the nanoscale, opening avenues for novel computing architectures.

5. Quantum Computing Concepts:

In recent years, quantum computing has emerged as a frontrunner in the realm of nanoelectronics. Harnessing the principles of quantum mechanics, quantum computers promise unprecedented computational power for solving complex problems.

This chapter embarks on a historical expedition, unveiling the evolution and milestones that propelled nanoelectronics from conceptual infancy to the cusp of quantum computing breakthroughs. As we traverse this journey, the quantum realm beckons, promising a future where computing defies the constraints of classical physics.

Navigating the Nanoscale Wonders: Unveiling the Power of Nanoelectronic Components

In the intricate domain of nanoelectronics, the miniaturization of electronic components unlocks a realm of extraordinary possibilities. Nanoscale electronic components redefine the boundaries of performance, efficiency and computational capabilities. Let's delve into the nanowonders of transistors, nanowires, nanotubes, and quantum dots.

1. Nanoscale Transistors:

At the forefront of nanoelectronics are nanoscale transistors, the backbone of electronic circuits. Miniaturizing transistors to the nanoscale brings about a quantum leap in performance. As these transistors shrink, electronic devices become faster, smaller, and more energy-efficient. The utilization of materials like carbon nanotubes and graphene paves the way for unprecedented advancements in microchip technology, leading to the creation of more powerful and compact devices.

2. Nanowires and Nanotubes:

Nanowires and nanotubes emerge as the architectural marvels, serving as the building blocks for nanoelectronic devices. Nanowires, resembling threads at the nanoscale, and nanotubes, cylindrical structures with remarkable conductive properties, facilitate the construction of intricate circuits and connections.

Their unique properties enable the development of smaller and more complex electronic components, pushing the boundaries of what is achievable in device engineering.

3. Quantum Dots:

Quantum dots, the enchanting entities within the nanoscale universe, harness the principles of quantum mechanics to revolutionize computing. These nanoscale semiconductor particles exhibit size-dependent electronic behaviour. Quantum dots possess the potential to encode quantum bits (qubits), leveraging quantum entanglement and superposition. The use of quantum dots in computing heralds a new era, where computation occurs at speeds and complexities unimaginable with classical computers.

As we navigate the nano revolution, these components collectively redefine the landscape of electronic devices, offering enhanced performance, miniaturization, and the harnessing of quantum properties for computing. The future unfolds with promises of smaller, more efficient, and incredibly powerful electronic devices, setting the stage for a transformative era in technology.

Nanoelectronic Marvels: Unveiling the Power Within

Nanoelectronic devices, operating at the frontier of technology, showcase an array of revolutionary components that redefine the capabilities of electronic systems. Let's embark on a journey into the intricate world of nanoscale memory technologies, quantum-level logic gates, and the transformative role of nanoelectronics in sensing and control.

1. Nanoscale Memory Technologies:

In the realm of memory devices, nanoelectronics introduces a paradigm shift with nanoscale memory technologies. These advancements enable the storage of information at an unprecedented scale. Nanoscale memory devices leverage the unique properties of nanomaterials, such as phase-change materials and magnetic nanoparticles. This not only enhances storage density but also facilitates faster data access, laying the foundation for the next generation of high-performance memory systems.

2. Quantum-Level Logic Gates:

At the heart of nanoelectronic computation lie quantum-level logic gates, transcending classical computing constraints. These gates harness the principles of quantum mechanics to process information in ways unimaginable in traditional systems. Quantum bits (qubits) replace classical bits, existing in a superposition of states. This enables parallel processing and intricate computations, promising a leap in computational power that could revolutionize fields like cryptography and optimization algorithms.

3. Nanoelectronics in Sensing and Control:

In the domain of sensing and control, nanoelectronics plays a pivotal role in creating highly sensitive and responsive systems. Nanoscale sensors and actuators utilize the unique properties of nanomaterials to detect and respond to stimuli with unparalleled precision. From medical applications to environmental monitoring, nanoelectronics enables devices that can adapt and interact with their surroundings at a scale previously thought impossible.

As we unveil the power within nanoelectronic devices, these innovations collectively redefine the landscape of computing, memory storage, and interactive systems. The nanoelectronic marvels promise a future where electronic devices are not just smaller and faster but also more adaptive and intelligent.

Beyond Bits: Nanoelectronic Marvels in Quantum and Neuromorphic Realms

In the ever-evolving landscape of nanoelectronics, two computational paradigms stand out, promising revolutionary advancements: quantum computing and neuromorphic computing.

1. Quantum Computing:

Quantum computing represents a paradigm shift from classical computing, leveraging the principles of quantum mechanics to perform computations at speeds unimaginable with classical bits. In classical computing, bits exist in either a 0 or 1 state. Quantum bits, or qubits, exist in a superposition of both states simultaneously. This enables parallel processing and the potential to solve complex problems exponentially faster than classical counterparts. Harnessing quantum states, such as entanglement and superposition, quantum computing holds promise for applications like optimization problems, cryptography, and simulations that surpass classical computing capabilities.

2. Neuromorphic Computing:

On the other frontier of nanoelectronics is neuromorphic computing, inspired by the human brain's architecture and functionality. This paradigm aims to emulate the brain's parallel processing and adaptive learning capabilities at the nanoscale. Neuromorphic systems use artificial neurons and synapses to mimic the brain's interconnected network. By integrating nanoscale components, neuromorphic computing brings about energy-efficient and highly parallel computing. The emulation of biological neural networks allows for tasks such as pattern recognition, learning, and decision-making. This paradigm holds immense potential for advancing artificial intelligence and cognitive computing.

As we venture into the nanoelectronic future, the fusion of quantum and neuromorphic computing opens doors to unprecedented computational power and intelligence. Beyond the limitations of classical computing, these paradigms herald a new era, promising breakthroughs in solving complex problems, understanding human cognition, and transforming the landscape of technology.

NanoRevolution: Transforming Computing with Nanoelectronics

Nanoelectronics is ushering in a new era of computing, offering breakthrough applications in high-performance computing, embedded systems, and paving the way for exciting future developments.

1. High-Performance Computing (HPC):

In the realm of supercomputing, nanoelectronics plays a pivotal role in pushing the boundaries of computational power. The integration of nanoscale components enhances processing speeds, reduces energy consumption and enables the development of exascale and quantum computers. Nanoelectronics in supercomputing holds the promise of solving complex problems, from simulations of molecular interactions to climate modelling, at unprecedented speeds.

2. Embedded Systems:

The integration of nanoelectronics into everyday devices is reshaping the landscape of technology through embedded systems. These systems, found in smartphones, smart appliances, and IoT devices, leverage nanoscale components to enhance performance, reduce size, and increase energy efficiency. Nanoelectronics in embedded systems empowers devices to process data locally, reducing dependence on cloud computing and enabling real-time, responsive applications.

3. Future Trends and Possibilities:

Anticipated developments in nanoelectronics hold the key to transformative advancements. Emerging hardware technologies, such as neuromorphic computing and quantum computing based on nanoscale devices, promise to revolutionize how we process information. Neuromorphic computing mimics the human brain's architecture, enabling advanced AI applications, while quantum computing harnesses the unique properties of quantum states for unparalleled computational capabilities.

The convergence of nanoelectronics with other cutting-edge technologies opens possibilities for developing energy-efficient, highly parallel computing systems. As we look to the future, nanoelectronics will continue to be a driving force behind innovations that reshape the computing landscape, offering solutions to complex challenges and unlocking unprecedented computational potentials.

NanoElectronics: Navigating Challenges and Embracing Opportunities

Nano electronics, at the intersection of nanotechnology and electronics, presents a realm of exciting possibilities coupled with unique challenges.

1. Manufacturing Challenges:
Precision and Scalability

Manufacturing at the nanoscale introduces precision challenges. Achieving atomic-level precision in constructing nanoelectronic components demands advanced fabrication techniques. Scalability is another hurdle, as traditional manufacturing methods struggle to maintain efficiency when working with nanoscale materials. Innovations in nanolithography and self-assembly techniques are crucial for overcoming these challenges.

Material Selection and Integration

Selecting suitable materials at the nanoscale is complex. Material properties drastically change at this level, impacting conductivity, durability, and overall functionality. Integrating these materials into functional devices without compromising their unique properties poses a significant challenge.

2. Ethical Considerations:
Societal Implications

The ethical landscape of nanoelectronics encompasses societal implications. Concerns arise about potential job displacement due to automation, privacy invasion through advanced surveillance technologies, and the societal divide caused by unequal access to nanotechnological advancements.

Environmental Impact

Nanoelectronics bring forth concerns about environmental sustainability. The manufacturing processes and disposal of nanoscale materials may have unforeseen consequences on ecosystems. Addressing these concerns involves developing eco-friendly manufacturing techniques and thorough lifecycle assessments.

Dual-Use Dilemma

The dual-use dilemma involves the potential misuse of nanoelectronics for both beneficial and harmful purposes. Applications in healthcare and energy may coexist with military applications, raising questions about responsible development and usage.

3. Opportunities:
Healthcare Innovations

Nano electronics offer transformative opportunities in healthcare, from targeted drug delivery systems to wearable health monitors. These advancements promise personalized medical treatments and early disease detection.

Energy Efficiency

Nanoelectronic devices can revolutionize energy storage and conversion. Improved efficiency in solar cells, energy harvesting, and energy storage devices could pave the way for sustainable energy solutions.

Balancing the challenges and opportunities in nanoelectronics requires collaborative efforts from researchers, policymakers, and industry stakeholders. Striking a harmonious balance between innovation and ethical considerations is essential for realizing the full potential of this groundbreaking technology.

Navigating Nanoelectronics: Real-World Triumphs and Lessons

In the realm of nanoelectronics, case studies illuminate the practical impact of groundbreaking technology on computing. These real-world implementations offer success stories and invaluable lessons for the industry.

1. Nanoelectronics in Cloud Computing
Success Story: Enhanced Performance

Real-world case studies showcase the integration of nanoelectronics in cloud computing infrastructure. This implementation has led to unparalleled improvements in processing speed and data handling capabilities, enabling seamless scalability and enhanced performance.

Lesson Learned: Scalability Challenges

Despite success, challenges in scaling nanoelectronic components persist. Lessons learned highlight the importance of addressing scalability issues to fully harness the potential of nanoelectronics in cloud computing environments.

2. Nanoelectronics in Life Sciences
Success Story: Advanced Characterization

Nanoelectronics has played a crucial role in life sciences, contributing to advanced nanomaterial characterization projects. These studies have provided deep insights into sample properties, aiding the development of cutting-edge technologies.

Lesson Learned: Comprehensive Characterization

Success in life sciences emphasizes the necessity for comprehensive nanomaterial characterization. The lesson learned underscores the importance of thorough evaluation for effective integration into diverse applications.

3. Nanoelectronics in High-Performance Computing
Success Story: Supercomputing Breakthroughs

Real-world implementations highlight nanoelectronics' impact on supercomputing, leading to breakthroughs in computational capabilities. Success stories demonstrate unparalleled processing speeds and energy efficiency in high-performance computing.

Lesson Learned: Power Consumption

The success in high-performance computing emphasizes the critical consideration of power consumption. Lessons learned underscore the need for sustainable and energy-efficient solutions in nanoelectronic applications.

These case studies unveil the triumphs and challenges of implementing nanoelectronics in diverse fields. Success stories underscore the transformative potential of nanoelectronics, while lessons learned guide the industry toward addressing scalability, comprehensive characterization, and sustainable energy consumption for a future shaped by nanotechnology.

Nanoelectronics: Illuminating the Future

The future of nanoelectronics holds extraordinary promise, poised for groundbreaking advancements and seamless integration with cutting-edge technologies.

1. Advancements on the Horizon
Potential Breakthroughs

Nanoelectronics is on the cusp of revolutionary breakthroughs. Innovations in materials science and fabrication techniques are expected to usher in unprecedented levels of precision and scalability. Quantum dots, 2D materials, and novel semiconductor designs are some of the key areas fuelling these advancements.

Energy Efficiency and Miniaturization

Anticipated breakthroughs include ultra-low-power nanodevices and the miniaturization of circuits to atomic scales. These developments not only promise enhanced performance but also contribute to the sustainability of electronic devices.

2. Integration with Other Technologies
Synergies with AI

The convergence of nanoelectronics and artificial intelligence (AI) opens new frontiers. Nanoscale devices can facilitate efficient neural network processing, enabling AI applications with unprecedented speed and energy efficiency.

IoT Connectivity

Nanoelectronics is set to play a pivotal role in the Internet of Things (IoT). Ultra-small sensors and devices, enabled by nanoelectronics, will seamlessly integrate into the IoT ecosystem, creating a network of interconnected and intelligent devices.

3. Quantum Communication

The marriage of nanoelectronics and quantum communication promises unparalleled secure communication channels. Nanoscale components can support the development of quantum processors and communication devices, paving the way for quantum computing and secure quantum networks.

In summary, the future prospects of nanoelectronics are luminous, with advancements poised to redefine the landscape of electronic devices. The seamless integration with AI, IoT, and quantum communication heralds an era of interconnected, intelligent, and secure technologies.

Nanoelectronics: Pioneering the Future of Computing

In conclusion, the trajectory of nanoelectronics is charting an exhilarating course toward transforming the landscape of computing. The insights garnered from recent explorations and studies shed light on the pivotal role nanoelectronics will play in shaping the future.

Key Insights Recapitulation
1. Precision at the Nanoscale

One of the fundamental insights lies in the ability of nanoelectronics to operate at the nanoscale. This precision allows for the creation of incredibly small yet powerful components, contributing to the development of ultra-compact and efficient electronic devices.

2. Integration with AI and Quantum Computing

The fusion of nanoelectronics with artificial intelligence and quantum computing emerges as a transformative revelation. Nanoscale devices facilitate the seamless integration of AI algorithms, enhancing computing capabilities and paving the way for quantum leaps in processing power.

3. Advancements in Device Engineering

Nanotechnological advancements have spurred progress in device engineering, resulting in more capable and compact computers. The impact is tangible, exemplified by the impressive capabilities of modern smartphones.

Significance of Nanoelectronics in the Future
1. Unprecedented Performance

The significance of nanoelectronics lies in its capacity to redefine performance benchmarks. As nanoscale components become ubiquitous, computing devices are poised to deliver unparalleled speed, energy efficiency, and computational prowess.

2. Sustainability and Miniaturization

Nanoelectronics contributes to the sustainability of computing by enabling the development of ultra-low-power devices. Simultaneously, the miniaturization of circuits to atomic scales ensures the creation of environmentally friendly and efficient electronic systems.

Nanoelectronics, with its remarkable precision and transformative integration capabilities, stands as the linchpin in the evolution of computing. It not only promises unprecedented technological achievements but also heralds a future where computing power knows no bounds.

7. Nanomedicine: Healing at the Molecular Level

Unveiling Nanomedicine: A Journey into the Microscopic Realm of Healing

Nanomedicine, a revolutionary field at the intersection of nanotechnology and medicine, holds the promise of transforming healthcare on a molecular level. Defined as the application of nanotechnology to medicine, nanomedicine involves manipulating materials and devices at the nanoscale to diagnose, treat, and prevent diseases. This convergence of disciplines represents a paradigm shift in the way we approach healthcare, offering unprecedented opportunities for precise and targeted interventions.

Historical Context and Evolution

The roots of nanomedicine trace back to the visionary concept of physicist Richard Feynman in his 1959 lecture, "There's Plenty of Room at the Bottom," where he envisaged the manipulation of individual atoms to create new materials and devices. However, it wasn't until the advent of nanotechnology in the late 20th century that the idea took concrete shape. The 1990s witnessed the formal inception of nanomedicine as a distinct field. Pioneering efforts by researchers like Eric Drexler fuelled the imagination with the possibility of nanoscale machines navigating the human body, repairing damaged cells, and precisely delivering therapeutic payloads. As advancements in nanotechnology accelerated, nanomedicine transitioned from a theoretical concept to a realm of tangible discoveries and applications.

In the early 21st century, breakthroughs in nanomedicine started permeating various medical domains. Targeted drug delivery, diagnostic imaging with nanoparticles and nanoscale biosensors became realities with the potential to revolutionize diagnostics and treatments. The introduction of nanoscale materials opened new vistas for designing more efficient and less invasive medical interventions. The evolution of nanomedicine is characterized by a synergy between the physical, chemical, biological, and digital worlds. Researchers and practitioners collaborate across disciplines, leveraging nanotechnology's precision to address medical challenges at an unprecedented scale. As nanomedicine continues to progress, it stands poised to redefine the landscape of healthcare, offering personalized, minimally invasive, and highly effective solutions for improved patient outcomes.

Unveiling Nanotechnology's Diagnostic Marvels: Precision in the Palm of Your Hand

Nanotechnology has emerged as a game-changer in diagnostics, offering unparalleled precision and sensitivity in disease detection. Leveraging nanoscale materials, imaging techniques, biosensors, and nanoprobes have all undergone transformative advancements, redefining the landscape of medical diagnostics.

Molecular Imaging for Disease Detection

Molecular imaging, a revolutionary nanotechnology application, allows for the visualization of cellular and molecular processes within the body. Techniques like positron emission tomography

(PET) and magnetic resonance imaging (MRI) have been enhanced with nanoscale contrast agents. These agents, often nanoparticles coated with specific molecules, enable targeted imaging of disease markers. This molecular precision not only facilitates early disease detection but also provides valuable insights into the underlying biological processes.

Biosensors and Nanoprobes: A Diagnostic Symphony

The integration of nanotechnology into biosensors has propelled diagnostic accuracy to unprecedented heights. Nanoscale materials, such as nanowires and quantum dots, serve as sensitive components in biosensors, enabling the real-time detection of biomolecules indicative of diseases. The synergy between nanotechnology and biosensors has led to the development of portable and point-of-care diagnostic devices, empowering healthcare professionals with rapid and efficient tools for on-the-spot diagnosis. Nanoprobes, engineered nanoscale particles with unique optical and magnetic properties, contribute significantly to diagnostic accuracy. They can be designed to target specific cells or biomarkers, providing enhanced contrast in imaging techniques like fluorescence imaging. This level of specificity allows for precise localization of abnormalities, aiding in the characterization and staging of diseases.

Advancements in Diagnostic Accuracy

The marriage of nanotechnology with diagnostics has ushered in an era of heightened accuracy and reduced invasiveness. Nanoparticles coated with ligands can selectively bind to disease markers, amplifying detection signals. Additionally, the use of nanomaterials in biosensors enhances sensitivity, allowing for the detection of trace amounts of biomolecules, even in early disease stages.

In summary, nanotechnology in diagnostics is not just a technological leap; it is a revolution in healthcare. The marriage of molecular imaging, biosensors, and nanoprobes has opened new frontiers, promising earlier and more accurate diagnoses, ultimately improving patient outcomes.

Nanomedicine Marvels: Precision, Efficacy, and Molecular Therapies

Nanomedicine is a revolutionary field at the intersection of nanotechnology and medicine and it has therefore ushered in a new era of therapeutic applications, offering precise interventions at the molecular level. Three key pillars, targeted drug delivery, nanoparticles in chemotherapy, and gene therapy, showcase the transformative potential of nanomedicine.

Targeted Drug Delivery: Precision Medicine Approaches

One of the hallmark features of nanomedicine is its ability to deliver therapeutic agents with pinpoint accuracy. Targeted drug delivery employs nanoparticles as carriers for drugs, ensuring they reach specific cells or tissues while sparing healthy ones. Surface modifications on nanoparticles enable them to recognize and bind to specific biomarkers on diseased cells, allowing for a highly targeted and personalized approach. This precision minimizes off-target effects, enhancing therapeutic efficacy.

Nanoparticles in Chemotherapy: Maximizing Efficacy, Minimizing Side Effects

Nanoparticles play a pivotal role in redefining chemotherapy by addressing its limitations. Conventional chemotherapy often lacks specificity, leading to systemic toxicity and adverse effects.

Nanoparticles, on the other hand, can encapsulate chemotherapy drugs, improving their solubility and stability. Moreover, these nanoparticles can passively accumulate in tumour tissues due to the enhanced permeability and retention (EPR) effect, increasing drug concentration at the target site. This selective accumulation enhances the therapeutic efficacy of chemotherapy while minimizing damage to healthy tissues, mitigating side effects.

Gene Therapy: Molecular-Level Manipulation

In the realm of nanomedicine, gene therapy takes centre stage, offering the ability to manipulate genes at the molecular level. Nanoparticles serve as delivery vehicles for genetic material, ensuring its safe and targeted delivery to cells. This approach allows for the correction of genetic defects, regulation of gene expression, and even the introduction of therapeutic genes. Gene therapy holds immense promise in treating genetic disorders, cancers, and other diseases at their root cause.

In summary, nanomedicine's therapeutic applications represent a paradigm shift in medicine, offering precision, enhanced efficacy, and molecular-level interventions that hold the potential to revolutionize healthcare.

Nanotechnology's Healing Touch: Revitalizing Regenerative Medicine

In the realm of regenerative medicine, nanotechnology emerges as a game-changer, offering innovative solutions to accelerate healing processes and enhance cellular regeneration. Two key facets, nanomaterials for tissue engineering and stem cell nanotechnology, showcase the transformative impact of nanotechnology in regenerative medicine.

Nanomaterials for Tissue Engineering: Accelerating Healing Processes

Nanomaterials play a pivotal role in tissue engineering, leveraging their unique properties to create scaffolds that mimic the extracellular matrix. These engineered scaffolds provide a conducive environment for cell adhesion, proliferation and differentiation. The nanoscale features of these materials, such as surface topography and controlled release of bioactive molecules, accelerate the healing processes. By guiding cellular behaviour at the molecular level, nanomaterials enhance tissue regeneration, offering a promising avenue for repairing damaged tissues and organs.

Stem Cell Nanotechnology: Advancements in Cellular Regeneration

Stem cell nanotechnology revolutionizes regenerative medicine by harnessing the potential of stem cells for cellular regeneration. Nanotechnology facilitates the precise manipulation and delivery of stem cells, ensuring their targeted integration into damaged tissues. Functionalized nanoparticles act as carriers for stem cells, enhancing their viability and therapeutic efficacy. Additionally, nanotechnology enables the modulation of the stem cell microenvironment, influencing their behaviour and differentiation. This level of control at the nanoscale significantly advances the precision and success of stem cell-based regenerative therapies.

In summary, nanotechnology in regenerative medicine heralds a new era of healing possibilities. Nanomaterials empower tissue engineering with accelerated healing, while stem cell nanotechnology opens avenues for precise cellular regeneration. This synergy between nanotechnology and regenerative medicine holds immense promise in reshaping the landscape of healthcare and offering novel solutions for treating a variety of conditions.

Revolutionizing Healthcare: The Era of Personalized Nanomedicine

In the dynamic landscape of healthcare, personalized nanomedicine stands as a beacon of innovation, ushering in a new era of tailored and patient-specific treatments. This groundbreaking approach involves the customization of therapeutic interventions based on individual profiles, leveraging nanotechnology to enhance precision and efficacy.

Patient-Specific Treatment: Tailoring Therapies with Precision

Personalized nanomedicine delves into the intricacies of each patient's unique biology, allowing for the tailoring of therapies to their specific needs. By incorporating nanoscale carriers, treatment protocols can be optimized, ensuring targeted delivery of therapeutic agents. This precision minimizes side effects, enhances drug efficacy, and improves overall treatment outcomes. The approach extends beyond traditional medicine though and also considers genetic, environmental, and lifestyle factors to craft a bespoke healthcare strategy for each individual.

Challenges and Future Prospects: Customized Healthcare on the Horizon

While personalized nanomedicine holds immense promise, it is not without challenges. The complexities of genetic and microenvironmental variations pose hurdles in achieving universal success. Overcoming these challenges requires interdisciplinary collaboration and advancements in technology. Future prospects indicate a shift toward fully customized healthcare, where nanomedicine plays a pivotal role in preventive strategies and early disease detection. As technologies evolve, the integration of artificial intelligence and big data analytics is expected to refine personalized treatment plans further.

In summary, personalized nanomedicine represents a paradigm shift in healthcare, offering treatments as unique as the individuals they serve. While challenges persist, the trajectory toward customized healthcare is unmistakable. The amalgamation of nanotechnology, individualized treatment plans, and cutting-edge technologies heralds a future where healthcare is not just universal but uniquely tailored to each person's distinct medical profile.

NanoRevolution: Nanomedicine's Battle Against Infectious Diseases

In the relentless fight against infectious diseases, nanomedicine emerges as a game-changing ally, leveraging innovative strategies to combat pathogens and revolutionize immunization. This segment explores two key facets of nanomedicine in infectious diseases: nanoparticle-based antimicrobial strategies for targeting pathogens and the integration of nanotechnology in vaccine development.

Targeting Pathogens: Nanoparticle-Based Antimicrobial Strategies

Nanoparticle-based antimicrobial strategies represent a formidable arsenal in the pursuit of eradicating pathogens. Engineered nanoparticles, such as silver nanoparticles, possess intrinsic antimicrobial properties. These nanoscale warriors exhibit enhanced surface area and unique physicochemical characteristics, enabling them to interact with microbial structures. By disrupting the integrity of bacterial cell membranes or interfering with viral replication, nanoparticle-based antimicrobial agents showcase unprecedented efficacy. Additionally, the multifunctionality of these nanoparticles allows for tailored approaches, addressing a spectrum of infectious agents.

Vaccines and Nanotechnology: Advancements in Immunization

Nanotechnology has catalyzed groundbreaking advancements in vaccine development, ushering in a new era of precision immunization. Nano-sized carriers, such as liposomes or virus-like particles, facilitate targeted delivery of antigens, enhancing the immune response. These carriers can mimic the natural infection process, eliciting robust and long-lasting immunity. Furthermore, nanovaccines enable the incorporation of multiple antigens, broadening protection against diverse strains of infectious agents. The integration of nanotechnology in vaccines also addresses challenges like stability and controlled release. This ensures optimal vaccine storage and controlled antigen release, maximizing efficacy. Beyond conventional vaccines, nanomedicine explores innovative approaches such as mRNA vaccines, exemplified by the success of certain COVID-19 vaccines.

In summary, nanomedicine's prowess in targeting pathogens and advancing immunization heralds a new era in infectious disease management. The marriage of nanotechnology with medical science opens avenues for more effective, precise, and adaptable strategies in the ongoing battle against infectious diseases.

Navigating the NanoFrontier: Ethical and Regulatory Challenges in Nanomedicine

As nanomedicine propels healthcare into the future, it brings forth ethical considerations and regulatory issues that demand careful navigation. This segment delves into the pivotal aspects of ensuring the safe and ethical use of nanomedicine, striking a balance between scientific progress and responsible practices and navigating the complex regulatory landscape.

Balancing Progress with Responsible Practices

The rapid evolution of nanomedicine necessitates a conscious effort to balance scientific advancements with ethical considerations. Key ethical concerns include informed consent, risk communication and ensuring the autonomy of individuals participating in clinical trials. Transparency in communicating potential risks and benefits becomes paramount, fostering a culture of responsible innovation. Researchers and practitioners must actively engage in ethical discourse, anticipating and addressing societal concerns to foster trust in nanomedical innovations.

Regulatory Landscape: Navigating Legal and Ethical Frameworks

The regulatory landscape of nanomedicine is intricate, requiring a comprehensive understanding of legal and ethical frameworks. Governments and international bodies play a crucial role in establishing guidelines and standards to govern the development, testing, and deployment of nanomedical technologies. Regulatory agencies need to adapt swiftly to the dynamic nature of nanomedicine, ensuring that protocols align with emerging technologies. Stakeholder collaboration, including scientists, policymakers, and ethicists, becomes essential to create adaptive regulations that encourage innovation while safeguarding public health. Navigating this landscape involves robust pre-market assessment, post-market surveillance and continuous monitoring of nanomedical products. Periodic ethical reviews and engagement with the public contribute to refining regulatory frameworks in response to evolving ethical considerations and technological advancements.

In summary, the ethical considerations and regulatory issues in nanomedicine demand a harmonious blend of scientific progress and responsible practices. A vigilant and collaborative approach is essential in order to navigate the complex ethical landscape and regulatory frameworks, ensuring the promise of nanomedicine is realized ethically and safely.

NanoRevolution: Transformative Case Studies in Nanomedicine

Nanomedicine has ushered in groundbreaking advancements, demonstrated by impactful case studies that highlight both successes and the challenges overcome in real-world applications.

Successful Case Studies:
1. Doxil (Liposomal Doxorubicin):

Impact: Doxil, a liposomal formulation of the chemotherapeutic agent doxorubicin, showcased enhanced efficacy with reduced side effects, significantly improving cancer treatment outcomes.

Lesson Learned: Liposomal delivery systems can enhance drug effectiveness while minimizing adverse effects.

2. Abraxane (Nanoparticle Albumin-Bound Paclitaxel):

Impact: Utilizing albumin-bound paclitaxel nanoparticles, Abraxane demonstrated improved solubility and enhanced delivery of paclitaxel, resulting in effective cancer treatment.

Lesson Learned: Nanoparticle formulations can overcome limitations of poorly soluble drugs, improving bioavailability.

3. SPIONs in Cancer Theranostics:

Impact: Superparamagnetic Iron Oxide Nanoparticles (SPIONs) served as imaging agents and drug carriers in cancer theranostics, allowing simultaneous diagnosis and treatment.

Lesson Learned: Integration of imaging and therapeutic capabilities enhances precision in cancer management.

Challenges Overcome:
1. Biocompatibility and Toxicity Concerns:

Challenge: Ensuring nanomaterials' safety.

Overcoming: Rigorous material characterization and toxicity studies have become essential.

2. Clinical Translation Hurdles:

Challenge: Transitioning nanomedicine from labs to clinical settings.

Overcoming: Addressing tumour microenvironment complexities and optimizing drug delivery systems.

3. Regulatory Adaptation:

Challenge: Evolving regulations.

Overcoming: Collaborative efforts between researchers, policymakers and regulatory bodies for adaptive frameworks.

The success stories and challenges overcome in nanomedicine underscore its transformative potential, emphasizing the need for continued innovation and collaboration in order to harness its full impact.

NanoMedicine 2.0: Unveiling the Future Horizons

The future of nanomedicine holds exciting possibilities with anticipated breakthroughs and seamless integrations with cutting-edge technologies, propelling healthcare into a new era.

Emerging Technologies in Nanomedicine:
1. Nano-Robots for Targeted Drug Delivery:

Breakthrough: Nano-sized robots designed for precise drug delivery to targeted cells, minimizing side effects and enhancing therapeutic outcomes.

2. Nanoparticle Theranostics:

Breakthrough: Integrating diagnostic and therapeutic functions within nanoparticles, enabling real-time monitoring and personalized treatment strategies.

3. AI-Powered Nanomedicine Design:

Breakthrough: AI algorithms assisting in the design of nanomedicines, optimizing structures for enhanced efficacy and minimal toxicity.

Integration with Other Medical Fields:
1. AI for Personalized Treatment Plans:

Synergy: AI algorithms analysing patient data to tailor nanomedicine treatments based on individual profiles, optimizing therapeutic outcomes.

2. Robotics in Surgical Nanomedicine:

Synergy: Robotic-assisted procedures utilizing nanoscale tools for precision surgeries, minimizing invasiveness and accelerating recovery times.

3. IoT-Enabled Nanodevices:

Synergy: Integration of nanodevices with the Internet of Things (IoT) for real-time monitoring of patient responses, enabling timely adjustments in treatment plans.

The convergence of nanomedicine with emerging technologies and interdisciplinary collaborations is poised to revolutionize healthcare. As we step into NanoMedicine 2.0, the synergy of nanotechnology with AI, robotics, and IoT promises a future where personalized, precise, and efficient medical interventions become the norm.

Nanomedicine: Pioneering the Future of Healthcare

Concluding our discussion of our journey through nanomedicine, it's evident that this revolutionary field has not only unfolded remarkable advances but is set to reshape the landscape of healthcare as we know it.

Key Advances in Nanomedicine:
1. Precision Drug Delivery Systems:

Nanoscaled drug delivery systems have emerged as a beacon of hope, offering unparalleled precision in targeting diseased cells while minimizing side effects.

2. Theranostic Nanoparticles:

The integration of diagnostic and therapeutic functionalities within nanoparticles has paved the way for theranostic applications, allowing real-time monitoring and personalized treatment strategies.

3. AI-Optimized Nanomedicine Design:

Artificial intelligence is now a crucial ally in designing nanomedicines, optimizing structures for enhanced efficacy and safety.

Nanomedicine's Role in Shaping Healthcare:

Nanomedicine is not just a scientific pursuit; it's a transformative force poised to shape the future of healthcare:

1. Revolutionizing Drug Delivery:

The ability to precisely deliver therapeutic agents to targeted cells revolutionizes drug delivery, promising more effective treatments with fewer side effects.

2. Enhancing Diagnostics and Therapeutics:

Nanomedicine seamlessly integrates diagnostics and therapeutics, offering a holistic approach to patient care and ushering in an era of personalized medicine.

3. Optimizing Treatment Strategies:

AI-optimized nanomedicine design ensures that treatment strategies are not only effective but tailored to individual patient profiles, maximizing therapeutic outcomes.

As we close this chapter on nanomedicine, it's with anticipation and excitement for the paradigm shift it promises in healthcare. Nanomedicine is not just a chapter; it's the future script of a healthier, more precise, and patient-centric healthcare narrative.

8. Harnessing the Power of the Tiny: Nanotechnology Revolutionizing Energy

Introduction

Nanotechnology, the manipulation of matter at the nanoscale, has emerged as a revolutionary force reshaping the energy landscape. In this chapter, we delve into the transformative role of nanotechnology in energy generation, storage, and efficiency, elucidating its profound significance in driving sustainable development and meeting the growing global energy demand.

1. Energy Generation:

Nanotechnology presents a paradigm shift in energy generation, offering innovative solutions to harness renewable sources more efficiently. By leveraging nanomaterials' unique properties, such as quantum confinement effects and high surface area-to-volume ratios, researchers have developed advanced photovoltaic devices capable of converting sunlight into electricity with unprecedented efficiency. Additionally, nanotechnology enables the development of next-generation fuel cells, enhancing the efficiency of hydrogen conversion into clean energy.

2. Energy Storage:

The quest for efficient energy storage solutions finds a promising ally in nanotechnology. Nanomaterials, such as carbon nanotubes and graphene, exhibit exceptional properties ideal for enhancing battery performance. These materials enable the development of high-capacity lithium-ion batteries with faster charging rates, longer lifespan, and enhanced safety. Moreover, nanotechnology facilitates the creation of supercapacitors capable of storing and releasing energy rapidly, addressing the intermittency challenges of renewable energy sources like wind and solar.

3. Energy Efficiency:

Nanotechnology plays a pivotal role in optimizing energy utilization across various sectors. By incorporating nanomaterials into building materials, manufacturers can improve insulation, reduce heat loss, and enhance energy efficiency in infrastructures. Moreover, nanotechnology-driven advancements in industrial processes enable precise control over material properties, minimizing energy consumption and waste production.

Nanotechnology's significance in the modern energy landscape extends beyond technological advancements. It represents a catalyst for achieving sustainability goals and mitigating environmental impacts associated with conventional energy sources. Furthermore, nanotechnology holds immense potential for enhancing energy access and affordability, particularly in underserved regions lacking reliable electricity infrastructure.

As we navigate the complexities of the 21st-century energy transition, understanding and harnessing the capabilities of nanotechnology are essential. This chapter seeks to explore the synergies between nanotechnology and energy, offering insights into how nanoscience is revolutionizing the way we generate, store, and utilize energy resources to build a more sustainable and resilient future.

Harnessing Nanotechnology for Sustainable Energy

Nanotechnology has revolutionized energy generation across various fronts, from enhancing solar power efficiency to optimizing wind turbine performance and advancing nuclear energy.

1. Solar Energy:

Nanomaterials play a crucial role in improving the efficiency of photovoltaic cells, the cornerstone of solar energy systems. By incorporating nanomaterials such as quantum dots and nanowires into solar cells, researchers have achieved better light absorption and charge separation, leading to higher conversion efficiencies. These advancements enable solar panels to generate more electricity from sunlight, making solar energy a more viable and cost-effective option for powering homes and businesses.

2. Wind Energy:

Nanotechnology offers significant potential in improving the performance of wind turbines. Nanomaterials are utilized to develop lightweight and robust composite materials for turbine blades, enhancing their durability and efficiency. Moreover, advanced coatings containing nanoparticles reduce friction and improve aerodynamics, enabling wind turbines to operate more efficiently and capture more energy from the wind. These innovations contribute to the growth of wind energy as a clean and renewable source of electricity.

3. Advanced Nuclear Energy:

Nanoparticles hold promise for enhancing the efficiency and safety of nuclear power generation. In nuclear reactors, nanoparticles can serve as additives to improve the performance of coolant fluids, enhance heat transfer, and mitigate radiation damage to reactor components. Additionally, nanotechnology facilitates the development of advanced fuel fabrication techniques, such as nanoparticle dispersion fuels, which improve fuel burnup and reduce the risk of nuclear accidents. These advancements pave the way for safer, more efficient, and sustainable nuclear energy production.

By harnessing nanotechnology, we can unlock the full potential of renewable and nuclear energy sources, driving the transition towards a cleaner, more sustainable energy future. Nanotechnology's role in energy generation extends beyond incremental improvements; it represents a fundamental shift in our approach to harnessing and utilizing energy resources, offering innovative solutions to address the challenges of climate change and energy sustainability.

Powering the Future: Nanomaterials Revolutionize Energy Storage

Nanomaterials have emerged as key players in revolutionizing energy storage technologies, offering remarkable advancements in batteries, supercapacitors, and fuel cells.

1. Enhanced Battery Performance:

Nanomaterials, such as nanowires, carbon nanotubes, and graphene, are integrated into battery electrodes to enhance storage capacity and charge-discharge rates. These materials provide a high surface area for ion storage, enabling more efficient ion diffusion and increasing the battery's energy density. Additionally, nanocoatings improve electrode stability and prevent undesirable side reactions, prolonging battery lifespan and enhancing safety.

2. Rapid Energy Release with Supercapacitors:

Supercapacitors, also known as ultracapacitors, leverage nanomaterials like activated carbon and transition metal oxides to achieve rapid energy release and recharge rate. Nanoporous structures in these materials facilitate high surface area and swift ion mobility, enabling supercapacitors to store and release energy efficiently. Nanomaterial-based electrodes enhance power density and cycling stability, making supercapacitors ideal for applications requiring quick bursts of energy, such as regenerative braking in electric vehicles and smoothing power fluctuations in renewable energy systems.

3. Clean Energy Conversion in Fuel Cells:

Nanomaterials play a vital role in fuel cells by catalysing electrochemical reactions for clean energy conversion. Platinum nanoparticles, for instance, serve as efficient catalysts for hydrogen oxidation and oxygen reduction reactions at the fuel cell's anode and cathode, respectively. These nanoparticles increase reaction kinetics, reduce overpotential, and improve fuel cell efficiency. Additionally, nanostructured membranes enhance proton conductivity and minimize gas crossover, enhancing fuel cell performance and durability. With nanomaterials, fuel cells offer a sustainable alternative to traditional combustion-based power generation, producing electricity with zero greenhouse gas emissions.

Nanomaterials' unique properties, including high surface area, tunable morphology, and exceptional conductivity, are driving innovation in energy storage technologies. By leveraging these materials, researchers are advancing battery capacity, supercapacitor efficiency, and fuel cell performance, paving the way for a cleaner, more sustainable energy future.

Illuminating Efficiency: Nanotechnology's Role in Energy Optimization

Nanotechnology is revolutionizing energy efficiency through advancements in smart grids and energy-efficient lighting technologies.

1. Smart Grids:

Nanosensors are integral components of smart grids, enabling efficient energy distribution and management. These miniature sensors often made from nanomaterials like carbon nanotubes or quantum dots, monitor electricity flow, voltage levels, and grid stability in real-time. By providing accurate data at the micro-level, nanosensors facilitate dynamic grid adjustments, optimizing energy distribution and reducing transmission losses. Additionally, nanotechnology enhances energy storage within smart grids, enabling better integration of renewable energy sources like solar and wind power.

2. Energy-Efficient Lighting:

Nanomaterials play a pivotal role in advancing energy-efficient lighting, particularly in Light Emitting Diodes (LEDs). Quantum dots, for example, are semiconductor nanocrystals that improve LED efficiency by converting a broader spectrum of light into visible wavelengths. This allows for more precise colour rendering and reduced energy consumption compared to traditional lighting technologies. Additionally, nanocoatings on LED surfaces enhance light extraction efficiency, minimizing internal reflection and maximizing brightness. Moreover, nanotechnology facilitates the development of flexible and transparent LEDs, expanding their applications in architectural lighting and display technologies.

By integrating nanotechnology into smart grids and lighting systems, energy efficiency is significantly enhanced; leading to reduced energy consumption, lower costs, and a more sustainable future.

Nano-Innovations: Pioneering Sustainable Energy Solutions

Nanotechnology offers groundbreaking solutions for sustainable energy production and environmental conservation. Two key areas where nanotech plays a crucial role are hydrogen production and carbon capture and storage (CCS).

1. Hydrogen Production:

Nanocatalysts revolutionize hydrogen production, particularly in the generation of green hydrogen. These catalysts, often composed of noble metals or metal oxides at the nanoscale, significantly enhance the efficiency of water electrolysis and reforming processes, facilitating the generation of hydrogen from renewable sources like solar and wind energy. By accelerating chemical reactions and minimizing energy losses, nanocatalysts contribute to the scalability and cost-effectiveness of green hydrogen production, making it a viable alternative to fossil fuels.

2. Carbon Capture and Storage (CCS):

Nanomaterials offer innovative solutions for mitigating climate change by capturing and storing CO_2 emissions. Functionalized nanoparticles, such as metal-organic frameworks (MOFs) and graphene-based materials, possess high surface areas and tailored properties that enhance CO_2 adsorption and storage capacities. These nanomaterials can be integrated into CCS technologies deployed in industrial plants and power stations, capturing CO_2 before it is released into the atmosphere. Moreover, nanotech enables the conversion of captured CO_2 into valuable products, such as carbon-neutral fuels or building materials, contributing to a circular carbon economy.

By harnessing nanotechnology in hydrogen production and CCS, we can accelerate the transition towards a sustainable energy future, reducing greenhouse gas emissions and mitigating climate change effects.

Navigating the Nano Frontier: Challenges and Future Paths

Nanotechnology presents both immense opportunities and daunting challenges, particularly concerning its environmental impact and the future trajectory of emerging nanotechnologies in the energy sector.

1. Environmental Impact:

Balancing nanotech advancements with sustainability remains a critical challenge. While nanotechnology offers solutions for sustainable energy, water purification, and pollution remediation, concerns arise regarding the environmental risks associated with engineered nanoparticles. These particles can potentially accumulate in ecosystems, leading to unknown ecological consequences. Moreover, the energy-intensive processes involved in nanomaterial synthesis and fabrication may contribute to carbon emissions and resource depletion. Addressing these challenges requires comprehensive risk assessment, regulatory frameworks, and green nanotechnology practices to ensure the responsible development and deployment of nanotechnologies.

2. Emerging Nanotechnologies in the Energy Sector:

Nanotechnology holds immense promise for revolutionizing the energy landscape with innovative solutions. One such game-changer is the development of nanomaterials for advanced energy storage systems, such as lithium-sulphur batteries and supercapacitors. These nanomaterials offer higher energy densities, faster charging rates, and longer cycle lives, addressing the limitations of traditional energy storage technologies. Additionally, nanocatalysts play a crucial role in catalytic processes for hydrogen production, enabling efficient and sustainable methods for generating clean energy. Furthermore, nanomaterials facilitate the optimization of solar cells and the enhancement of photovoltaic efficiency, paving the way for cost-effective and scalable solar energy harvesting. These emerging nanotechnologies have the potential to significantly reduce carbon emissions, increase energy accessibility, and propel the transition towards a greener and more sustainable energy future.

Navigating the challenges and opportunities presented by nanotechnology requires a multidisciplinary approach, collaboration between stakeholders, and a steadfast commitment to sustainability principles. By addressing environmental concerns and harnessing the transformative power of emerging nanotechnologies, we can pave the way for a more sustainable and prosperous future for generations to come.

Unveiling the Power of Nanotechnology: Case Studies and Ongoing Research

Nanotechnology has made remarkable strides in various fields, with real-world case studies showcasing its successful implementations and ongoing research providing valuable insights for future advancements, particularly in the energy sector.

1. Successful Implementations:
Sunscreen:

Nanoparticles incorporated into sunscreens enhance their effectiveness by providing better UV protection without leaving a white residue, revolutionizing sun protection.

Renewable Energy:

Nanotechnology enhances renewable energy systems' efficiency. For instance, wind energy turbines can be made more efficient by using lighter and stronger nanomaterials, reducing maintenance costs and increasing energy output.

2. Lessons Learned and Ongoing Research:
Efficiency Optimization:

Ongoing research in nanotechnology energy projects focuses on improving the efficiency of renewable energy systems. By leveraging nanomaterials and nanodevices, researchers aim to overcome the limitations of traditional energy technologies, leading to more sustainable and cost-effective energy solutions.

Environmental Impact:

Lessons learned from nanotechnology energy projects emphasize the importance of assessing environmental impacts. Researchers recognize the need for sustainable practices and regulatory frameworks in order to mitigate potential risks associated with nanomaterials' use in energy applications.

Interdisciplinary Collaboration:

The interdisciplinary nature of nanotechnology research underscores the importance of collaboration between scientists, engineers, policymakers, and stakeholders. By fostering collaboration, researchers can leverage diverse expertise in order to address complex challenges and accelerate technological innovation.

These case studies and ongoing research highlight nanotechnology's transformative potential in addressing pressing energy challenges. By learning from past successes and ongoing endeavours, researchers can chart a course towards a more sustainable and energy-efficient future.

Navigating Ethical Frontiers: Regulation and Responsibility in Nanotechnology

Navigating the ethical and regulatory landscape in nanotechnology is crucial for ensuring responsible development, especially in the energy sector. Here's a comprehensive look at responsible nanotechnology in energy and the policy and regulations guiding its development:

1. Responsible Nanotechnology in Energy:
Safety Protocols:

Implementing rigorous safety protocols is essential to address concerns regarding the potential environmental and health impacts of nanomaterials used in energy applications. By conducting thorough risk assessments and adopting precautionary measures, such as containment strategies and exposure controls, researchers can mitigate potential hazards.

Sustainability:

Emphasizing sustainability is paramount in responsible nanotechnology for energy. This involves considering the entire lifecycle of nanoproducts, from raw material extraction to disposal, to minimize environmental footprint and ensure long-term viability.

2. Policy and Regulations:
Guiding Frameworks:

Robust regulatory frameworks are essential to guide nanotechnology development in the energy sector. Policies should address issues such as safety, environmental impact, ethical considerations, and intellectual property rights. By establishing clear guidelines, policymakers can foster innovation while safeguarding public welfare.

International Collaboration:

Given the global nature of nanotechnology, international collaboration is key to harmonizing regulatory standards and ensuring consistency across borders. Collaborative efforts facilitate information sharing, capacity building, and the development of common regulatory approaches, promoting responsible nanotechnology adoption worldwide.

In summary, navigating the ethical and regulatory landscape in nanotechnology requires a multifaceted approach encompassing responsible development practices, robust policy frameworks, and international cooperation. By prioritizing safety, sustainability, and collaboration, stakeholders can harness the transformative potential of nanotechnology in the energy sector while ensuring its ethical and regulatory compliance.

Unveiling the Future: Nanotechnology's Evolution in Energy Systems

Nanotechnology holds immense promise for shaping the future of energy systems through innovative solutions and collaborative endeavours. Here's a glimpse into the anticipated innovations and collaborative efforts in nanotech's role in future energy systems:

1. Anticipated Innovations and Breakthroughs:
Enhanced Energy Storage:

Nanotechnology is poised to revolutionize energy storage by enabling the development of advanced batteries with higher energy densities, faster charging rates, and longer lifespans. Innovations such as nanocomposite materials and nanostructured electrodes promise to address current limitations and accelerate the adoption of renewable energy sources.

Improved Solar Energy Conversion:

Nanomaterials offer unprecedented opportunities for enhancing the efficiency of solar cells and lowering the cost of solar energy production. Through precise control of nanoscale structures and novel light-trapping mechanisms, researchers aim to achieve higher conversion efficiencies and broader spectral absorption, paving the way for widespread deployment of solar power.

2. Collaborative Efforts and Global Initiatives:
International Research Consortia:

Collaborative research consortia bring together scientists, engineers, policymakers, and industry stakeholders from around the world to advance nanotechnology for sustainable energy. Initiatives like the International NanoScience Community promote knowledge exchange, joint projects, and shared resources to accelerate progress and address global energy challenges.

Public-Private Partnerships:

Public-private partnerships play a vital role in driving nanotechnology innovation in the energy sector. Collaborative initiatives between governments, research institutions, and industry leaders facilitate technology transfer, funding support, and commercialization efforts, fostering a conducive ecosystem for translating research discoveries into real-world applications.

In summary, the future of nanotechnology in energy systems is characterized by groundbreaking innovations and concerted global efforts. By harnessing nanotechnology's potential to revolutionize energy storage, solar energy conversion, and other critical aspects of the energy landscape, we can pave the way for a sustainable and resilient future.

Energizing Tomorrow: Nanotechnology's Sustainable Power

In conclusion, nanotechnology stands as a beacon of hope in the quest for sustainable energy solutions. Through its myriad applications, nanotechnology has revolutionized the energy landscape, offering innovative approaches to enhance efficiency, reduce environmental impact, and pave the way for a cleaner, greener future.

1. Recapitulation of Key Contributions:
Enhanced Energy Conversion:

Nanotechnology has propelled advancements in energy conversion technologies, such as solar cells and fuel cells, by enabling the development of high-efficiency, cost-effective materials and devices.

Improved Energy Storage:

Nanomaterials have played a pivotal role in enhancing energy storage systems, including batteries and supercapacitors, by increasing energy density, enhancing cycling stability, and reducing charging times.

Sustainable Energy Production:

Nanotechnology has facilitated the production of renewable energy sources, such as wind and solar power, by optimizing efficiency, reducing costs, and mitigating environmental impacts.

2. Affirming Nanotechnology's Impact on Energy Sustainability:

- Nanotechnology has significantly contributed to advancing energy sustainability by promoting the transition towards cleaner, more efficient energy systems.
- By leveraging nanotechnology, we can address pressing energy challenges, including climate change, resource depletion, and energy security concerns.
- The scalability, versatility, and affordability of nanotechnology make it a potent tool for driving sustainable energy innovation on a global scale.

So, in essence, nanotechnology holds immense promise for shaping the future of energy by unlocking new possibilities, pushing the boundaries of efficiency, and fostering sustainability. As we continue to harness the transformative power of nanotechnology, we must prioritize collaboration, innovation, and responsible stewardship to realize its full potential in creating a brighter, more sustainable energy future for generations to come.

9. Cultivating Tomorrow: Nanotechnology's Impact on Agriculture

Introduction

Nanotechnology has emerged as a groundbreaking frontier in agriculture, offering innovative solutions to address pressing challenges in food production, environmental sustainability, and crop protection. In this chapter, we explore the evolution and scope of nanofertilizers and nanopesticides, two pioneering applications that hold immense potential to revolutionize modern agriculture.

1. Evolution of Nanotechnology in Agriculture:

Nanotechnology's foray into agriculture marks a paradigm shift in farming practices, leveraging the manipulation of matter at the nanoscale in order to enhance crop yield, nutrient uptake, and pest management. The journey began with the advent of nanoscale materials and their exploration for agricultural applications, leading to the development of nanofertilizers and nanopesticides.

2. Scope of Nanofertilizers:

Nanofertilizers represent a novel approach to nutrient delivery, wherein essential elements are encapsulated within nanoscale carriers to improve their efficiency and availability to plants. These nanocarriers enable controlled release of nutrients, targeted delivery to specific plant tissues, and enhanced uptake through root systems. Moreover, nanofertilizers can mitigate nutrient leaching, minimize environmental impact, and promote sustainable farming practices.

3. Scope of Nanopesticides:

Nanopesticides offer a promising avenue for crop protection by harnessing nanotechnology to enhance the efficacy, stability, and targeted delivery of pest control agents. By encapsulating pesticides within nanocarriers, such as nanoparticles or nanocapsules, nanopesticides can achieve controlled release, prolonged activity, and reduced environmental contamination. Additionally, nanopesticides enable precise targeting of pests while minimizing exposure to non-target organisms and reducing pesticide residues in soil and water.

In summary, the evolution and scope of nanotechnology in agriculture underscore its transformative potential to address the complexities and demands of modern farming. As we delve deeper into the realm of nanofertilizers and nanopesticides, we uncover a wealth of opportunities to enhance agricultural productivity, sustainability, and resilience in the face of global challenges. Through interdisciplinary collaboration and innovative research, nanotechnology continues to pave the way for a more efficient, equitable, and environmentally conscious agricultural sector.

Nanofertilizers: Revolutionizing Nutrient Management

Nanofertilizers represent a cutting-edge approach to nutrient management in agriculture, leveraging nanotechnology to enhance nutrient absorption, improve efficiency, and minimize environmental impact. In this segment, we delve into the transformative potential of nanofertilizers and their key components: nanoparticles, controlled release systems, and nanoencapsulation techniques.

1. Nanoparticles in Fertilizers:

Nanoparticles, with their incredibly small size and high surface area-to-volume ratio, offer unique advantages in fertilizers. These nanoparticles can carry essential nutrients such as nitrogen, phosphorus, and potassium, facilitating their efficient uptake by plant roots. Moreover, nanoparticles can penetrate the root cell membrane more easily, enabling enhanced nutrient absorption and utilization by plants. This improved nutrient delivery system ensures that plants receive the required nutrients in optimal quantities, leading to healthier growth and higher yields.

2. Controlled Release Systems:

One of the key challenges in traditional fertilization methods is the rapid leaching of nutrients into the soil, leading to nutrient runoff and environmental pollution. Controlled release systems embedded within nanofertilizers address this issue by releasing nutrients gradually over an extended period. These systems utilize nanostructured matrices or coatings that regulate the release of nutrients in response to environmental factors such as soil moisture, temperature, and pH levels. By providing nutrients in a sustained manner, controlled release systems not only improve nutrient efficiency but also minimize environmental contamination, ultimately promoting sustainable agricultural practices.

3. Nanoencapsulation Techniques:

Nanoencapsulation involves encapsulating nutrients within nanoscale carriers or matrices to protect them from degradation and enhance their stability. These carriers can be composed of biodegradable polymers, lipids, or inorganic materials, providing a protective barrier against external factors such as moisture, UV radiation, and microbial activity. Additionally, nanoencapsulation enables the controlled release of nutrients, ensuring sustained uptake by plants over time. By safeguarding nutrients from premature degradation and facilitating their gradual release, nanoencapsulation techniques prolong the availability of nutrients in the soil and optimize their utilization by plants, contributing to improved crop productivity and reduced environmental impact.

In summary, nanofertilizers represent a revolutionary approach to nutrient management in agriculture, offering enhanced nutrient absorption, controlled release, and protection against environmental degradation. By harnessing the power of nanoparticles, controlled release systems, and nanoencapsulation techniques, nanofertilizers hold immense promise for sustainable farming practices, ensuring optimal nutrient delivery to plants while minimizing adverse environmental effects.

Nanopesticides: Precision Pest Management

Nanopesticides represent a groundbreaking approach to pest management, offering precision targeting, increased efficacy, and reduced environmental impact. In this segment, we explore the key features of nanopesticides, including targeted delivery systems, enhanced efficacy through nanoformulations, and the use of smart nanomaterials for timely pest control.

1. Targeted Delivery Systems:

Nanopesticides utilize targeted delivery systems to deliver active ingredients precisely to the intended pest organisms while minimizing exposure to non-target organisms and the environment.

These systems can be engineered to release pesticides selectively in response to specific cues, such as pest pheromones or environmental conditions. By ensuring precise delivery, targeted systems minimize environmental residue and reduce the risk of pesticide drift, thereby enhancing the safety and sustainability of pest management practices.

2. Increased Efficacy with Nanoformulations:

Nanoformulations of pesticides offer several advantages over conventional formulations, including increased solubility, stability, and bioavailability of active ingredients. Nano-sized particles enable better penetration and adhesion to pest surfaces, enhancing the efficacy of pesticides and reducing the required dosage. Moreover, nanoformulations can encapsulate multiple active ingredients or adjuvants, allowing for synergistic effects and improved pest control outcomes. By harnessing the power of nanotechnology, nanoformulations elevate the effectiveness of pesticides, leading to more sustainable pest management practices.

3. Smart Nanomaterials for Timely Action:

Smart nanomaterials employed in nanopesticides can respond to pest cues or environmental stimuli, enabling timely and targeted pest control interventions. These materials can be designed to release pesticides only when triggered by specific signals, such as changes in pH, temperature, or pest activity. By responding dynamically to pest presence or environmental conditions, smart nanomaterials ensure that pesticides are deployed precisely when and where they are needed, optimizing pest management efficacy while minimizing pesticide usage and environmental impact.

In summary, nanopesticides offer a paradigm shift in pest management by leveraging targeted delivery systems, nanoformulations for increased efficacy, and smart nanomaterials for responsive pest control. These advancements promise to revolutionize agriculture by enabling more precise, efficient, and sustainable pest management practices.

Nanotechnology's Harvest: Boosting Crop Yield and Nutrition

Nanotechnology has significantly impacted crop yield and quality by enhancing both productivity and nutritional value, revolutionizing agriculture.

1. Improved Crop Productivity:

Nanofertilizers have demonstrated remarkable efficacy in enhancing crop yield. For instance, in a study published in ScienceDirect, slow and controlled release nanofertilizers were shown to efficiently deliver nutrients, minimizing leaching and promoting sustainable crop growth. Additionally, nanoencapsulation techniques have enabled the targeted delivery of nutrients, improving nutrient uptake efficiency and minimizing nutrient losses during fertilization. These advancements have led to significant yield enhancements in various crops, ensuring food security and economic prosperity.

2. Nutrient Enrichment and Biofortification:

Nanotechnology-enabled biofortification strategies have emerged as powerful tools for enhancing the nutritional value of crops. By leveraging nanomaterials, researchers have successfully increased the micronutrient content in staple food crops, addressing micronutrient deficiencies prevalent in many regions. Agronomic biofortification methods, such as nanoparticle-mediated nutrient delivery, have been effective in enhancing the nutritional quality of crops without compromising yield. Furthermore, smart nanoparticles capable of responding to environmental cues have facilitated targeted nutrient delivery, ensuring optimal nutrient absorption by plants and improving human health outcomes.

Overall, nanotechnology's impact on crop yield and quality has been transformative, offering sustainable solutions to global agricultural challenges. Through improved crop productivity and nutrient enrichment techniques, nanotechnology has not only increased agricultural output but also contributed to addressing malnutrition and food insecurity worldwide.

Advancing Environmental Sustainability through Nanotechnology

Nanotechnology emerges as a pivotal tool in fostering environmental sustainability, offering innovative solutions in order to mitigate ecological degradation and to promote biodiversity conservation.

1. Reduced Environmental Footprint:

Nanotechnology plays a crucial role in minimizing environmental pollution and mitigating runoff in various sectors. Nano-enabled agrochemicals, such as nanofertilizers and nanopesticides, facilitate precise delivery of nutrients and pesticides to plants, reducing excess usage and limiting their leaching into water bodies. Additionally, nanoremediation strategies harness nanoparticles to degrade pollutants, offering targeted pollution control and remediation.

2. Enhanced Soil Health and Biodiversity:

Nanotechnology contributes to soil health maintenance and biodiversity conservation by improving agricultural practices. Nano-enhanced soil amendments enhance soil structure, water retention, and nutrient availability, fostering healthy soil ecosystems and promoting plant growth. Furthermore, nanosensors monitor soil quality, enabling precision agriculture practices that optimize resource utilization while minimizing environmental impact.

3. Balancing Agricultural Intensification with Conservation:

Nanotechnology aids in achieving a delicate balance between agricultural intensification and conservation efforts. Nanoformulations optimize crop productivity, allowing for increased yields on existing agricultural lands without further expansion into natural habitats. Moreover, nanomaterials improve crop resistance to pests and diseases, reducing the need for chemical interventions that can harm beneficial organisms and disrupt ecosystem balance.

By harnessing the power of nanotechnology, we can address pressing environmental challenges while fostering sustainable practices that safeguard ecosystems and promote the well-being of current and future generations.

Navigating Nanotechnology: Challenges and Considerations

Nanotechnology holds immense promise but also presents unique challenges and considerations that must be carefully addressed to maximize its benefits while minimizing potential risks.

1. Regulatory Frameworks:

Establishing robust regulatory frameworks is crucial to ensure the safe and ethical development and application of nanotechnology. These frameworks address safety concerns by mandating thorough risk assessments of nanomaterials and products. Additionally, ethical considerations, such as privacy, equity, and environmental impact, are integral components of regulatory guidelines to promote responsible innovation.

2. Adoption Barriers:

Despite its potential, nanotechnology faces several adoption barriers that hinder its widespread implementation. These barriers include:

a). *Technical Complexity:*

Nanotechnology involves intricate scientific principles and manufacturing processes, posing challenges for researchers and developers. Overcoming these technical hurdles requires significant expertise and investment in research and development.

b). *Cost:*

The high cost associated with nanotechnology research, development, and production presents a significant barrier, particularly for small and medium-sized enterprises (SMEs). Access to funding and resources is crucial to overcome financial barriers and facilitate innovation in nanotechnology.

c). *Regulatory Uncertainty:*

The lack of standardized regulations and guidelines for nanotechnology poses challenges for businesses and researchers.

Unclear regulatory requirements can lead to delays in product development and market entry, discouraging investment and innovation in the field.

d). Public Perception and Acceptance:

Public perception of nanotechnology, influenced by factors such as media coverage and societal values, can impact its acceptance and adoption. Addressing concerns about safety, ethical implications, and potential risks is essential to foster trust and acceptance among stakeholders.

Overcoming these adoption barriers requires collaborative efforts from governments, industry stakeholders, researchers, and policymakers. Strategies such as increasing investment in research and development, promoting interdisciplinary collaboration, and enhancing public awareness and education about nanotechnology can help address these challenges and accelerate its adoption. While nanotechnology offers tremendous potential to revolutionize various industries, including healthcare, electronics, and energy, it is essential to navigate the associated challenges and considerations effectively. By implementing robust regulatory frameworks, addressing adoption barriers, and fostering collaboration and innovation, we can harness the full potential of nanotechnology while ensuring its safe and ethical use.

Pioneering the Future: Innovations and Collaborations in Nanotechnology for Agriculture

Nanotechnology holds immense promise for revolutionizing agriculture, with emerging technologies poised to address key challenges and drive sustainable agricultural practices.

1. Emerging Nanotechnologies:

Anticipated advances in agricultural nanotechnology include:

a). Nanoencapsulation:

Nanoencapsulation enables the targeted delivery of nutrients, pesticides, and growth regulators to plants, minimizing wastage and maximizing efficacy. This technology enhances crop protection and nutrient uptake, contributing to improved yields and quality.

b). Nanosensors:

Nanosensors provide real-time monitoring of soil health, water quality, and plant physiological parameters, facilitating precision agriculture. By detecting early signs of stress or disease, nanosensors enable timely interventions, reducing losses and optimizing resource use.

c). Nano-biofertilizers:

Nano-biofertilizers enhance nutrient availability and uptake by plants, promoting sustainable soil fertility and reducing reliance on chemical fertilizers. These innovative fertilizers improve nutrient use efficiency, minimize environmental pollution, and support long-term soil health.

2. Collaborative Research and Global Initiatives:

Collaborative efforts and global initiatives play a pivotal role in shaping the future landscape of nanotechnology in agriculture:

1. International Research Consortia:

Multidisciplinary research consortia bring together scientists, policymakers, and industry stakeholders from across the globe to address common challenges and foster innovation in agricultural nanotechnology. By leveraging diverse expertise and resources, these consortia accelerate the translation of research into practical solutions.

2. Public-Private Partnerships:

Collaborations between public institutions and private companies drive the development and commercialization of nanotechnology-based agricultural products and technologies. These partnerships facilitate technology transfer, investment, and market access, promoting the adoption of innovative agricultural practices.

3. Global Policy Frameworks:

International organizations and initiatives, such as the Food and Agriculture Organization (FAO) and the Global Nanotechnology Network (GNN), establish policy frameworks and guidelines to govern the safe and responsible use of nanotechnology in agriculture. By promoting regulatory harmonization and knowledge exchange, these initiatives support sustainable development and ensure the ethical deployment of nanotechnologies.

In summary, the future of nanotechnology in agriculture is characterized by cutting-edge innovations and collaborative endeavours aimed at addressing global food security challenges and advancing sustainable agricultural practices. By harnessing the potential of emerging nanotechnologies and fostering international cooperation, we can pave the way for a more resilient, efficient, and environmentally friendly agricultural sector.

Nanotechnology's Harvest: Transforming Agriculture

Nanotechnology has ushered in a new era for agriculture, revolutionizing traditional practices and offering innovative solutions to age-old challenges. Through various case studies, we witness the tangible impact of nanotechnology on agricultural productivity, sustainability, and food security.

Successful Applications:

1. Enhanced Crop Yield:

Nanoparticle-based fertilizers and pesticides have demonstrated remarkable efficacy in enhancing crop yield and resilience. These nanoformulations deliver nutrients and agrochemicals precisely to plants, minimizing wastage and environmental contamination.

2. Soil Remediation:

Nanomaterials such as nanoscale zero-valent iron (nZVI) hold promise for soil remediation, effectively removing heavy metals and pollutants from contaminated soils. This application not only restores soil health but also mitigates the adverse effects of industrial pollution on agricultural lands.

3. Precision Agriculture:

Nanosensors enable real-time monitoring of soil moisture, nutrient levels, and crop health, empowering farmers with data-driven insights for precise irrigation and fertilization. This targeted approach optimizes resource use and minimizes environmental impact.

Lessons Learned and Ongoing Research:
1. Risk Assessment:

Despite the promising benefits, concerns persist regarding the potential toxicity of nanomaterials to the environment and human health. Ongoing research focuses on comprehensive risk assessment protocols to ensure the safe deployment of nanotechnology in agriculture.

2. Ecosystem Interactions:

Understanding the interactions between nanoparticles and soil microorganisms, plants, and beneficial insects is crucial for predicting their long-term environmental impacts. Lessons from ongoing studies shed light on the complex dynamics within agricultural ecosystems and guide sustainable nanotechnology practices.

3. Regulatory Frameworks:

Developing robust regulatory frameworks is imperative to govern the use of nanotechnology in agriculture. Lessons learned from early adoption cases emphasize the importance of proactive regulation to address potential risks and ensure responsible innovation.

In summary, nanotechnology offers a promising pathway to address the pressing challenges facing agriculture in the 21st century. While successful applications demonstrate its potential to revolutionize farming practices and enhance food production, ongoing research underscores the importance of cautious optimism and responsible stewardship. By embracing lessons learned and fostering interdisciplinary collaboration, we can harness the transformative power of nanotechnology to cultivate a sustainable and resilient agricultural future.

Revolutionizing Agriculture: The Nanotechnology Impact

Nanotechnology has emerged as a game-changer in agriculture, offering innovative solutions to address the challenges faced by the industry. Through the application of nanomaterials and nanodevices, agriculture has witnessed remarkable advancements that have significantly contributed to sustainability and productivity. One of the key contributions of nanotechnology in agriculture is in crop protection and enhancement. Nanopesticides and nano-fertilizers enable targeted delivery of agrochemicals, minimizing environmental contamination and maximizing nutrient absorption by plants. This not only improves crop yield but also reduces the ecological footprint of farming practices.

Moreover, nanotechnology plays a crucial role in soil remediation and improvement. Nanomaterials can help in restoring soil quality, enhancing its fertility, and mitigating soil degradation issues. By promoting soil health, nanotechnology contributes to sustainable agriculture practices that support long-term food production without compromising natural resources. Additionally, nanotechnology facilitates precision agriculture through sensors and nanodevices that monitor soil conditions, crop growth, and environmental factors in real-time. This data-driven approach enables farmers to make informed decisions, optimize resource utilization, and minimize waste, ultimately leading to more efficient and sustainable farming practices.

In summary, nanotechnology has revolutionized agriculture by offering novel solutions to age-old challenges. Its impact extends beyond conventional farming methods, fostering sustainability, resilience, and productivity in agricultural systems. As we continue to harness the potential of nanotechnology, the future of agriculture looks promising, with the prospect of feeding a growing global population while safeguarding the planet's resources.

10. Environmental Implications of Nanotechnology

Unveiling Nano: Exploring Environmental Impacts

The proliferation of nanotechnology has been rapid and far-reaching, permeating various different sectors with its transformative potential. In medicine, nanoparticles are revolutionizing drug delivery systems, enabling targeted therapy with reduced side effects. In electronics, nanomaterials facilitate the development of smaller, faster, and more efficient devices, powering the evolution of consumer electronics. Moreover, in energy, nanotechnology offers promising avenues for sustainable solutions, from enhancing solar cell efficiency to improving battery performance. However, alongside these remarkable advancements, nanotechnology brings forth profound environmental implications that necessitate careful consideration. As nanoparticles become increasingly prevalent in consumer products and industrial processes, concerns about their impact on ecosystems and human health emerge. The unique properties of nanoparticles, such as their high surface area and reactivity, raise questions about their potential to accumulate in the environment and interact with living organisms in unforeseen ways. Understanding the environmental implications of nanotechnology requires a multidisciplinary approach, integrating insights from environmental science, toxicology, risk assessment, and policy analysis. Researchers strive to evaluate the fate and transport of nanoparticles in various environmental compartments, assess their toxicity to organisms, and develop strategies to mitigate potential risks.

In this chapter, we embark on a journey to explore the environmental impacts of nanotechnology. We delve into the complexities of nanoparticle interactions with environmental matrices, ranging from soil and water to air and biota. Through case studies and research findings, we unravel both the beneficial applications and potential hazards of nanotechnology in environmental contexts. By elucidating the intricate interplay between nanomaterials and ecosystems, we aim to inform decision-makers, stakeholders, and the public about the challenges and opportunities associated with harnessing nanotechnology for a sustainable future. Nanotechnology holds immense promise to address pressing environmental issues, from pollution remediation to resource conservation. Yet, it is essential to navigate this nascent field with caution, ensuring that technological advancements align with environmental stewardship and human well-being. By fostering dialogue, collaboration, and responsible innovation, we can harness the transformative potential of nanotechnology while safeguarding the delicate balance of our planet's ecosystems.

Nanoparticles in the Environment: Sources and Mobility

Nanoparticles in the environment originate from various sources, including industrial activities, consumer products, and natural processes. Understanding these sources and their dispersion mechanisms is crucial for assessing their environmental impact.

Sources of Nanoparticles/Nanowaste:
1. Industrial Activities:

Manufacturing processes, such as nanomaterial synthesis and fabrication, release nanoparticles into the environment. These include nanoparticles used in electronics, cosmetics, and medical devices.

2. Consumer Products:

Everyday products like sunscreens, clothing, and food packaging contain engineered nanoparticles. Through use and disposal, these nanoparticles can enter the environment via wastewater or landfill leachate.

3. Natural Processes:

Natural sources, such as volcanic eruptions and weathering of rocks, also contribute to nanoparticle release. These nanoparticles may become airborne or enter aquatic systems through erosion and sediment transport.

Dispersion and Mobility:
1. Airborne Transport:

Nanoparticles can disperse over long distances through atmospheric transport. Ultrafine particles can remain suspended in the air for extended periods, facilitating their transport to remote areas.

2. Waterborne Transport:

Nanoparticles can enter water bodies through runoff, wastewater discharge, and atmospheric deposition. Once in water, they may undergo sedimentation or remain suspended, depending on their size and surface properties.

3. Soil Interaction:

Nanoparticles deposited onto soil surfaces may interact with soil minerals and organic matter, influencing their mobility and fate. Factors such as soil texture and pH affect nanoparticle retention and transport in the soil matrix.

4. Biological Uptake:

Nanoparticles can be taken up by plants, animals, and microorganisms, leading to potential ecological impacts. This uptake pathway influences nanoparticle mobility and bioaccumulation within food webs.

Understanding the sources and mobility of nanoparticles in the environment is essential for assessing their potential risks to ecosystems and human health. Continued research into nanoparticle fate and transport mechanisms is necessary to develop effective mitigation strategies and regulatory measures.

Nanotechnology: Navigating Toxicity and Ecological Impact

Nanotechnology offers groundbreaking advancements, but its potential toxicity and ecological impact raise significant concerns across various fronts.

Ecotoxicological Concerns:
1. Effects on Flora and Fauna:

Nanoparticles can accumulate in ecosystems, affecting plant growth and soil microbial communities. Aquatic organisms like fish and algae are vulnerable to nanoparticle toxicity, disrupting ecological balance.

Human Health Implications:
1. Occupational Hazards:

Workers involved in nanomaterial production face exposure risks, leading to respiratory issues and skin irritation. Proper safety measures and protective equipment are essential to mitigate these risks.

2. Public Exposure:

Nanoparticles in consumer products can inadvertently expose the public to potential health risks. Further research is needed to understand the long-term effects of chronic nanoparticle exposure.

Regulatory Landscape:
1. Current Frameworks:

Regulatory agencies worldwide are developing guidelines for nanomaterial safety assessment and risk management. However, challenges persist in standardizing testing protocols and addressing emerging nanomaterials.

2. Challenges:

Limited data on nanoparticle toxicity and diverse nanomaterial properties complicate regulatory efforts. Harmonizing toxicity data and establishing robust risk assessment frameworks remain critical challenges.

Navigating the toxicity and ecological impact of nanotechnology requires a multifaceted approach, encompassing rigorous research, enhanced safety measures, and effective regulatory frameworks. Collaboration between scientists, policymakers, and industry stakeholders is vital to ensure the responsible development and deployment of nanotechnology.

Revolutionizing Environmental Monitoring with Nanotechnology

Nanotechnology has revolutionized environmental monitoring through the development of nanosensors and innovative detection technologies, enhancing water and air quality surveillance.

Nanosensors Advancements:
1. Increased Sensitivity:

Nanosensors offer heightened sensitivity, enabling the detection of minute quantities of pollutants and contaminants in environmental samples.

2. Miniaturization:

Nanotechnology facilitates the miniaturization of sensors, allowing for the deployment of compact, portable monitoring devices capable of real-time data collection in various environmental settings.

3. Specificity:

Advanced nanosensors exhibit high specificity, distinguishing between different types of pollutants with precision and accuracy.

Water and Air Quality Monitoring Applications:
1. Pollution Surveillance:

Nanotechnology enhances pollution surveillance by enabling the detection and quantification of contaminants such as heavy metals, nitrates, pathogens, and volatile organic compounds (VOCs) in water and air.

2. Remote Monitoring:

Nanosensors facilitate remote monitoring of water bodies and air quality, providing real-time data on pollutant levels and environmental conditions.

3. Early Warning Systems:

Nanotechnology enables the development of early warning systems for environmental hazards, alerting authorities and communities to potential threats to water and air quality.

4. Precision Agriculture:

Nanosensors contribute to precision agriculture by monitoring soil moisture, nutrient levels, and pesticide concentrations, optimizing resource utilization and minimizing environmental impacts.

Nanotechnology's integration into environmental monitoring holds promise for enhancing our understanding of ecosystem health, facilitating timely interventions to mitigate pollution, and safeguarding public health and natural resources.

Nanomaterial Remediation and Green Nanotech Solutions for Sustainable Environmental Cleanup

Nanomaterials offer promising avenues for environmental cleanup, with innovative strategies targeting soil and water contamination. Additionally, green nanotech approaches provide sustainable solutions to mitigate pollution and preserve ecological balance.

Remediation and Treatment Strategies:
1. Nanoscale Remediation Techniques:

Nanotechnology-based remediation techniques, such as nanosensors and nanoremediation, target pollutants at the molecular level, offering enhanced efficiency and precision in soil and water cleanup.

2. Soil Cleanup:

Nanomaterials, including nano zero-valent iron (nZVI) and carbon-based nanomaterials, facilitate soil remediation by sequestering heavy metals, pesticides, and organic contaminants, thus restoring soil fertility and productivity.

3. Water Purification:

Nanotechnology-enabled water treatment methods, such as membrane filtration and photocatalysis, remove pollutants like heavy metals, organic compounds, and pathogens, ensuring access to clean drinking water and preserving aquatic ecosystems.

Green Nanotech Approaches:
1. Sustainable Nanomaterial Synthesis:

Green nanotechnology emphasizes eco-friendly synthesis methods using renewable resources and environmentally benign processes, reducing the environmental footprint of nanomaterial production.

2. Biodegradable Nanomaterials:

Development of biodegradable nanomaterials ensures minimal environmental impact by facilitating degradation and reducing long-term persistence in ecosystems.

3. Phytoremediation with Nanoparticles:

Integration of nanomaterials in phytoremediation processes enhances plant-based remediation of contaminated sites, improving soil quality and promoting sustainable land use practices.

By harnessing the potential of nanotechnology and adopting green nanotech approaches, we can address environmental challenges effectively while promoting sustainability and ecological resilience.

Sustainable Nanotechnology: Building a Greener Future

Sustainable nanotechnology involves critical considerations to minimize environmental impact and promote eco-friendly practices throughout the life cycle of nanomaterials. Two key aspects include life cycle assessment and the design of green nanomaterials.

Life Cycle Assessment (LCA):
1. Evaluation of Environmental Footprints:

LCA assesses the environmental impact of nanomaterials from raw material extraction to disposal, identifying potential hotspots and guiding improvements.

2. Holistic Approach:

LCA considers various factors such as resource depletion, energy consumption, emissions, and waste generation, providing a comprehensive understanding of nanotechnology's environmental implications.

Designing Green Nanomaterials:
1. Sustainability Principles:

Green nanomaterials are designed with principles like atom economy, renewable resources, non-toxicity, and biodegradability, ensuring minimal environmental harm throughout their life cycle.

2. Renewable Resources:

Utilizing renewable resources for nanomaterial synthesis reduces reliance on finite resources and minimizes ecological footprint.

3. Biodegradability and Non-Toxicity:

Green nanomaterials are engineered to degrade into harmless by-products and exhibit low toxicity, mitigating potential risks to human health and the environment.

By integrating life cycle assessment practices and embracing green nanomaterial design principles, sustainable nanotechnology strives to advance technological innovation while safeguarding environmental and human well-being.

Unveiling the Unintended: Nanotechnology's Environmental Consequences

Nanotechnology, while promising numerous benefits, also poses unintended consequences for the environment, stemming from nanomaterial persistence and emerging risks.

Nanomaterial Persistence:

1. Long-Term Effects:

Nanomaterials can persist in the environment for extended periods, potentially leading to chronic exposure and cumulative impacts on ecosystems.

2. Accumulation:

Nanoparticles may accumulate in soil, water, and organisms over time, exacerbating environmental contamination and posing risks to biodiversity.

Emerging Risks:
1. Unforeseen Environmental Impact:

The novelty of nanotechnology introduces uncertainty regarding its ecological repercussions, as unforeseen interactions with ecosystems and organisms may occur.

2. Complexity of Effects:

Nanomaterials' intricate properties may trigger unexpected environmental effects, challenging traditional risk assessment methodologies and necessitating adaptive management strategies.

Anticipating and mitigating these unintended consequences require interdisciplinary research efforts, enhanced regulatory frameworks, and proactive risk management strategies.

Pioneering the Path: Future Trajectories of Nanotechnology in Environmental Science

Nanotechnology promises groundbreaking solutions for environmental challenges, but future directions and research needs are crucial for its responsible implementation.

Knowledge Gaps:

1. Nanomaterial Fate and Transport:

Understanding the behaviour of nanomaterials in diverse environmental matrices, including soil, water, and air, remains a priority.

2. Ecotoxicological Impacts:

Assessing the long-term effects of nanomaterial exposure on ecosystems and biodiversity is essential for risk management.

3. Sustainable Synthesis Methods:

Developing eco-friendly approaches for nanomaterial synthesis to minimize environmental footprint.

Collaborative Initiatives:
1. International Collaboration:

Collaborative efforts like the International Symposium on Nanotechnology and Occupational Health foster global dialogues on responsible nanotechnology.

2. Research Consortia:

Organizations like the National Nanotechnology Initiative (NNI) facilitate interdisciplinary research collaborations to address environmental implications.

3. Public-Private Partnerships:

Collaborative ventures between academia, industry, and government promote the development of sustainable nanotechnologies.

By focusing on these research areas and fostering collaborative endeavours, the field of nanotechnology can advance towards environmentally sustainable solutions, ensuring a harmonious coexistence between technology and nature.

Striking a Balance: Navigating Nanotechnology's Environmental Impact

In conclusion, the environmental implications of nanotechnology present a complex landscape that requires careful navigation. While nanotechnology holds tremendous potential for addressing environmental challenges, it also poses significant risks that must be addressed to ensure responsible development and deployment. Throughout this chapter, we have explored the dual nature of nanotechnology's impact on the environment. On one hand, nanomaterials offer innovative solutions for pollution remediation, water treatment, and sustainable energy production. However, their widespread use raises concerns about unintended environmental consequences, such as ecotoxicity and nanomaterial persistence. To address these challenges, it is imperative to adopt a precautionary approach that prioritizes environmental protection while fostering technological advancement. This includes robust risk assessment protocols, stringent regulatory frameworks, and transparent communication channels. Collaborative efforts between governments, industries, and research institutions are essential to ensure that nanotechnology evolves sustainably.

Moving forward, it is essential to continue researching the environmental fate and transport of nanomaterials, ecotoxicological impacts, and sustainable synthesis methods. By addressing these knowledge gaps, we can better understand the potential risks associated with nanotechnology and develop strategies to mitigate them. Ultimately, the goal is to strike a balance between harnessing the transformative power of nanotechnology and safeguarding the environment for future generations. This requires a holistic approach that integrates scientific expertise, ethical considerations, and stakeholder engagement. The environmental implications of nanotechnology underscore the importance of responsible innovation. By proactively addressing environmental considerations and promoting sustainability, we can unlock the full potential of nanotechnology while minimizing its negative impact on the environment.

11. Ethical and Regulatory Dimensions

Navigating the Ethical and Regulatory Landscape of Nanotechnology

Nanotechnology, the manipulation of matter on an atomic or molecular scale to create materials with unique properties, has emerged as a transformative force with far-reaching implications across various sectors. From healthcare and electronics to energy and environmental sustainability, nanotechnology promises groundbreaking advancements that could revolutionize industries and improve human life. Over the past few decades, nanotechnology has witnessed exponential growth and diversification, driven by advances in science, engineering, and technology. This rapid expansion of nanotechnology has led to a proliferation of nanomaterials and nanoproducts in the market, ranging from consumer goods to cutting-edge medical devices. Nanotechnology's impact on various industries is undeniable, with applications ranging from drug delivery systems and biomedical implants to efficient energy storage devices and environmental remediation technologies.

However, along with its promising potential, nanotechnology also raises significant ethical and regulatory concerns that must be addressed to ensure its responsible development and safe deployment. As nanomaterials interact with biological systems, questions regarding their toxicity, environmental impact, and long-term health effects have emerged. Ethical considerations surrounding nanotechnology encompass a wide range of issues, including equity in access to nanotechnology-based innovations, privacy concerns related to nanoscale surveillance technologies, and the potential for unintended consequences, such as environmental disruption and societal inequalities. Moreover, regulatory frameworks governing the development and commercialization of nanotechnology products are still evolving, presenting challenges in ensuring adequate oversight and risk management.

In this chapter, we will delve into the ethical and regulatory dimensions of nanotechnology, exploring the complex interplay between technological innovation, societal values, and public policy. By examining key issues and emerging trends, we aim to provide insights into the ethical dilemmas and regulatory challenges inherent in the development and deployment of nanotechnology. Through critical analysis and informed dialogue, we seek to foster a deeper understanding of the ethical implications of nanotechnology and facilitate the development of responsible governance frameworks that promote innovation while safeguarding human health and environmental well-being.

Addressing Ethical Challenges in Nanotechnology

Nanotechnology, while promising remarkable advancements, brings forth a host of ethical dilemmas that demand careful consideration. These challenges span various domains, including privacy and security, equity and access, environmental and health impacts, and the dual-use dilemma.

1. Privacy and Security Concerns:

Nanotechnology's integration into everyday products and systems raises concerns about individual privacy and data security. Nanoscale sensors and devices embedded in consumer goods or medical implants may collect sensitive personal information, leading to potential breaches and surveillance.

2. Equity and Access:

Ensuring equitable distribution of benefits and access to nanotechnology advancements is crucial. There's a risk that only affluent individuals or developed regions may benefit from nanotech innovations, exacerbating social inequalities. Efforts are needed to promote inclusive access and address disparities in resource allocation and technology diffusion.

3. Environmental and Health Impacts:

Nanomaterials' unique properties may pose unforeseen risks to ecosystems and human health. Nanoparticles' small size and high reactivity raise concerns about their potential toxicity and environmental persistence. Understanding and mitigating these impacts are essential to ensure responsible development and deployment of nanotechnology.

4. The Dual-Use Dilemma:

Nanotechnological advancements often have dual-purpose applications, with both beneficial and harmful potential. This dilemma arises when technologies developed for beneficial purposes are repurposed for malicious intent, such as nanomaterials used for warfare or surveillance. Balancing innovation with ethical considerations requires proactive measures to prevent misuse while fostering beneficial applications.

Addressing these ethical challenges requires a multifaceted approach involving policymakers, researchers, industry stakeholders, and civil society. Regulatory frameworks must be established to safeguard privacy, ensure equitable access, and mitigate environmental and health risks. Additionally, promoting transparency, accountability, and ethical education can foster a culture of responsible innovation in nanotechnology. Moreover, international collaboration and interdisciplinary dialogue are essential to address ethical challenges comprehensively. By fostering global cooperation and sharing best practices, the nanotechnology community can navigate ethical dilemmas while maximizing the societal benefits of technological innovation.

In summary, while nanotechnology holds immense potential for transformative change, it also presents significant ethical challenges. By proactively addressing privacy and security concerns, promoting equity and access, mitigating environmental and health impacts, and managing the dual-use dilemma, stakeholders can ensure that nanotechnology contributes positively to society while minimizing potential risks.

Regulatory Frameworks in Nanotechnology

Nanotechnology's rapid advancement necessitates robust regulatory frameworks to ensure its safe and responsible development. This segment explores key components of regulatory efforts, including the U.S. National Nanotechnology Initiative (NNI), international collaboration, and ethical guidelines.

1. U.S. National Nanotechnology Initiative (NNI):

The NNI coordinates nanotechnology research and development across various U.S. federal agencies, including the National Science Foundation (NSF), the National Institutes of Health (NIH), and the Environmental Protection Agency (EPA). The primary U.S. regulatory body overseeing nanotechnology-related matters is the Food and Drug Administration (FDA), responsible for regulating nanotechnology products in areas such as healthcare, food, and cosmetics. Additionally, the Occupational Safety and Health Administration (OSHA) addresses workplace safety concerns related to nanomaterial exposure.

2. International Regulatory Collaboration:

International cooperation and harmonization of regulatory approaches are crucial for effectively managing nanotechnology's global impact. Organizations like the Organisation for Economic Co-operation and Development (OECD) facilitate dialogue and information exchange among member countries to develop common principles and guidelines for nanotechnology regulation. Efforts such as the OECD Working Party on Manufactured Nanomaterials aim to establish regulatory frameworks that address safety, risk assessment, and responsible innovation.

3. Ethical Guidelines and Standards:

Ethical considerations are integral to nanotechnology research and application. Establishing ethical guidelines and standards helps guide responsible conduct and decision-making in nanotech-related activities. International cooperation and global regulatory standards play a vital role in unifying research approaches and evaluating ethical implications. Ethical frameworks address issues such as privacy, security, equity, and the dual-use dilemma, aiming to balance technological innovation with societal welfare. Regulatory frameworks and ethical guidelines are dynamic and evolving, adapting to emerging technologies and societal needs. Continuous dialogue among stakeholders, including policymakers, scientists, industry, and civil society, is essential to address regulatory gaps, promote transparency, and ensure public trust in nanotechnology.

In summary, regulatory frameworks in nanotechnology encompass national initiatives like the NNI, international collaboration facilitated by organizations such as the OECD, and ethical guidelines aimed at guiding responsible conduct and decision-making. These efforts are vital for fostering innovation while safeguarding human health, environmental integrity, and societal well-being in the rapidly evolving field of nanotechnology.

Ethical Oversight in Nanotechnology

Ethical oversight is paramount in nanotechnology in order to ensure responsible and safe advancement. This segment explores two critical aspects of ethical oversight: research ethics and human trials with informed consent.

1. Research Ethics:

Nanotechnology research must adhere to stringent ethical standards in order to mitigate potential risks and uphold integrity. Key considerations include:

a). Safety:

Researchers must prioritize the safety of both humans and the environment throughout all stages of nanotech research, from development to application. This involves identifying and mitigating potential hazards associated with nanomaterials.

b). Transparency:

Ethical research practices demand transparency in reporting methods, results, and any conflicts of interest. Open communication fosters trust among stakeholders and allows for independent scrutiny of findings.

c). Accountability:

Researchers are accountable for the societal implications of their work. Ethical reflection should guide decision-making, particularly in areas with ambiguous risks or uncertain outcomes.

2. Human Trials and Informed Consent:

Ethical conduct in human trials involving nanotechnology requires rigorous adherence to principles of autonomy, beneficence, and justice. Key considerations include:

a). Informed Consent:

Participants in nanotech human trials must provide informed consent voluntarily, with a clear understanding of the study's purpose, procedures, potential risks, and benefits. Informed consent ensures respect for individual autonomy and promotes trust between researchers and participants.

b). Risk-Benefit Assessment:

Ethical oversight necessitates a thorough evaluation of the risks and potential benefits associated with nanotech interventions. Researchers must weigh the potential harms against the expected benefits to ensure that the trial's conduct aligns with principles of beneficence.

c). Vulnerability:

Special attention must be given to vulnerable populations participating in nanotech trials, such as children, the elderly, and individuals with diminished capacity. Extra safeguards may be required to protect their rights and well-being.

Ethical oversight in nanotechnology is a dynamic process that requires continuous evaluation and adaptation to emerging challenges and advancements. Collaboration between researchers, regulatory bodies, ethicists, and the broader community is essential to establish and uphold ethical standards that promote the responsible development and application of nanotechnology.

Future Considerations in Nanotechnology

As nanotechnology continues to advance, it is crucial to anticipate future ethical challenges and regulatory responses to ensure responsible innovation and mitigate potential risks. This segment explores key considerations in both areas:

1. Anticipating Ethical Challenges:

As nanotechnological developments progress, several ethical challenges may arise, including:

a). Privacy and Security:

With the increasing use of nanoscale sensors and devices, protecting individual privacy and data security becomes paramount. Ethical dilemmas may arise concerning the collection, storage, and use of personal data obtained through nanotechnology-enabled applications.

b). Equity and Access:

Ensuring equitable access to nanotechnology benefits and addressing potential disparities in access among different socioeconomic groups will be critical. Ethical considerations may arise regarding the distribution of resources and opportunities associated with nanotech advancements.

c). Environmental and Health Impacts:

As nanomaterials become more prevalent in various industries, ethical concerns regarding their potential environmental and health impacts will intensify. Addressing issues such as nanoparticle toxicity, ecological disruption, and unintended consequences will be imperative.

d). Dual-Use Dilemma:

Nanotechnology's dual-purpose nature, with both beneficial and potentially harmful applications, poses ethical dilemmas. Balancing the promotion of beneficial uses with measures to prevent misuse or unintended harm will require careful consideration.

2. Evolving Regulatory Responses:

Regulatory frameworks must evolve to keep pace with technological advancements in nanotech. Key considerations include:

a). Risk-Based Regulation:

Regulatory agencies need to adopt a risk-based approach to assess and manage the potential risks associated with nanotechnology. This involves evaluating both the benefits and risks of nanomaterials and products to inform regulatory decisions.

b). Interdisciplinary Collaboration:

Addressing the complex challenges posed by nanotechnology requires interdisciplinary collaboration among scientists, policymakers, ethicists, and stakeholders. Regulatory responses should incorporate diverse perspectives to develop effective and inclusive governance frameworks.

c). Adaptive Governance:

Regulatory frameworks should be adaptable and responsive to emerging scientific evidence and technological developments. Flexibility in regulations allows for timely adjustments to address new risks and opportunities in nanotechnology.

d). International Cooperation:

Given the global nature of nanotechnology, international cooperation and harmonization of regulatory standards are essential. Collaborative efforts among nations can promote consistency, facilitate information exchange, and enhance the effectiveness of regulatory oversight.

By proactively addressing ethical challenges and adapting regulatory responses, stakeholders can foster the responsible development and safe deployment of nanotechnology, unlocking its vast potential while safeguarding against potential risks.

Case Studies in Nanotechnology Ethics

Nanotechnology has presented various ethical dilemmas throughout its development, but it has also seen instances where ethical regulations effectively addressed concerns. Here are some case studies illustrating historical ethical dilemmas and successful ethical regulation:

1. Historical Ethical Dilemmas:
a). *Privacy and Security:*

Early in nanotechnology's development, concerns emerged regarding the potential misuse of nanoscale materials for surveillance or espionage purposes. The ability to create ultra-small sensors raised ethical questions about privacy invasion and data security.

b). *Environmental Impact:*

The release of engineered nanoparticles into the environment raised ethical concerns about their potential ecological impact. Questions arose regarding the long-term consequences of nanoparticles on ecosystems and biodiversity.

c). *Health and Safety:*

As nanomaterials found their way into consumer products and medical applications, ethical dilemmas emerged regarding their potential health risks. Concerns included nanoparticle toxicity, bioaccumulation, and unintended health consequences.

2. Success Stories in Ethical Regulation:
a). *Risk Assessment and Management:*

Regulatory agencies have developed frameworks for risk assessment and management to address potential health and environmental risks associated with nanotechnology. By conducting thorough risk assessments, regulatory bodies can identify potential hazards and implement appropriate control measures.

b). *Interdisciplinary Collaboration:*

Ethical regulation in nanotechnology benefits from interdisciplinary collaboration between scientists, ethicists, policymakers, and stakeholders. By incorporating diverse perspectives, regulatory frameworks can address complex ethical dilemmas and promote responsible innovation.

c). *Public Engagement and Transparency:*

Successful ethical regulation involves engaging the public in decision-making processes and ensuring transparency in regulatory decisions. By fostering public trust and participation, regulatory bodies can better address societal concerns and enhance the legitimacy of regulatory frameworks.

These case studies demonstrate the importance of proactive ethical consideration and effective regulatory oversight in guiding the development and deployment of nanotechnology. By learning from past ethical dilemmas and implementing successful regulatory strategies, stakeholders can navigate the complex ethical landscape of nanotechnology while maximizing its benefits and minimizing potential risks.

Conclusion: Balancing Innovation and Responsibility in Nanotechnology

Nanotechnology holds immense potential to revolutionize various fields, from medicine to energy and beyond. However, with great innovation comes the responsibility to ensure that ethical considerations and regulatory frameworks are in place to guide its development and application. Throughout this chapter, we have explored the multifaceted dimensions of ethics and regulation in nanotech, recognizing both the challenges and successes in navigating this complex landscape.

Ethical considerations in nanotechnology are paramount due to its profound implications for society, the environment, and human health. Issues such as privacy, security, environmental impact, and equitable access to benefits must be carefully addressed in order to mitigate potential risks and maximize societal benefits. Furthermore, the interdisciplinary nature of nanotechnology requires collaboration between scientists, ethicists, policymakers, and stakeholders in order to ensure that ethical principles are integrated into all stages of research and development.

As we conclude this chapter, it is imperative to emphasize the call to action for continuous improvement in ethical and regulatory practices in nanotechnology. The rapid pace of technological advancement demands adaptive and forward-thinking regulatory frameworks that can effectively address emerging ethical challenges. This requires ongoing dialogue and engagement among stakeholders in order to identify potential risks, anticipate future ethical dilemmas, and develop proactive strategies to address them. Transparency and public engagement are essential in order to build trust and to ensure that regulatory decisions align with societal values and preferences.

In summary, while nanotechnology offers unprecedented opportunities for innovation and advancement, it also poses significant ethical and regulatory challenges that must be navigated responsibly. By striking a balance between innovation and responsibility, we can harness the full potential of nanotechnology while safeguarding the well-being of individuals and the planet. The journey towards ethical excellence in nanotech requires a commitment to continuous learning, collaboration, and adaptation in order to ensure that our actions today pave the way for a sustainable and ethically responsible future.

12. Future Prospects and Emerging Trends

Exploring the Future of Nanotechnology: Evolution, History, and Emerging Trends

Nanotechnology, the science of manipulating matter at the nanoscale, has undergone a remarkable evolution since its inception, paving the way for transformative advancements across various industries. As we embark on a journey to explore the future prospects and emerging trends in nanotech, it is crucial to understand its evolutionary trajectory and significant milestones. Throughout the late 20th and early 21st centuries, nanotechnology experienced exponential growth, driven by groundbreaking discoveries and technological advancements. Key milestones include the development of carbon nanotubes, which exhibit remarkable mechanical, electrical, and thermal properties, and the discovery of fullerenes, ushering in a new era of nanomaterials. These advancements propelled nanotechnology into the mainstream, fuelling innovations in electronics, medicine, energy, and environmental remediation.

The purpose of this chapter is to delve into the future prospects and emerging trends in nanotech, building upon the rich history and evolution of this field. By examining past achievements and current developments, we aim to forecast the trajectory of nanotechnology and anticipate the transformative impact it will have on society and the economy. As we set the stage for exploration, it is essential to highlight the multifaceted nature of nanotechnology's future prospects. From nanomedicine and personalized healthcare to nanoelectronics and quantum computing, the potential applications of nanotech are vast and diverse. Emerging trends such as the convergence of nanotechnology with artificial intelligence, biotechnology, and materials science present unprecedented opportunities for innovation and disruption.

Furthermore, this chapter aims to address the societal and ethical implications of nanotechnology's proliferation, emphasizing the importance of responsible innovation and equitable access to its benefits. By critically examining both the promises and perils of nanotech, we can chart a course towards a future where technological advancements are leveraged to enhance human well-being and address global challenges. This chapter will serve as a roadmap for navigating the evolving landscape of nanotechnology. By understanding its evolution, historical advancements, and current trends, we can anticipate the future trajectories of nanotech and harness its transformative potential for the betterment of society and the environment.

Anticipated Developments in Nanotechnology: Quantum Computing, Nanomedicine, and Energy Solutions

Nanotechnology continues to drive innovation across various fields, with anticipated developments poised to revolutionize quantum computing, nanomedicine, and energy harvesting and storage. These advancements hold immense promise for addressing pressing challenges and improving the quality of life for individuals worldwide.

1. Quantum Computing:

Nanotechnology plays a pivotal role in advancing quantum computing capabilities by enabling the development of quantum processors and qubits, the fundamental building blocks of quantum computers. Anticipated developments include:

a). Qubit Stability:

Nanoscale materials such as superconducting nanowires and semiconductor quantum dots offer enhanced qubit stability, prolonging coherence times and improving computational efficiency.

b). Quantum Dot Computing:

Quantum dots, semiconductor nanoparticles with quantum mechanical properties, hold promise for creating scalable quantum computing platforms. Their small size and tunable properties make them ideal candidates for qubit implementation.

2. Nanomedicine:

Innovations in nanotechnology are anticipated to drive significant breakthroughs in medical applications, shaping the future of healthcare delivery and disease treatment. Key developments include:

a). Targeted Drug Delivery:

Nanoparticles engineered with precise targeting capabilities can deliver therapeutics directly to diseased cells, minimizing systemic side effects and enhancing treatment efficacy.

b). Diagnostic Nanosensors:

Nanoscale sensors capable of detecting biomarkers at ultra-low concentrations offer early disease detection and real-time monitoring, revolutionizing diagnostic practices and improving patient outcomes.

3. Energy Harvesting and Storage:

Nanotechnology holds immense potential for transforming energy harvesting and storage technologies, paving the way for sustainable and efficient energy solutions. Anticipated developments include:

a). Nanomaterials for Solar Cells:

Nanoscale materials such as quantum dots and perovskite nanocrystals enable the development of next-generation solar cells with higher efficiency and lower production costs.

b). Nanocomposite Energy Storage:

Nanocomposite materials incorporating nanoscale additives enhance the performance and durability of energy storage devices such as batteries and supercapacitors, enabling longer lifetimes and faster charging rates.

In summary, anticipated developments in nanotechnology hold immense promise for advancing quantum computing, nanomedicine, and energy solutions. By harnessing the unique properties of nanoscale materials and structures, researchers and innovators are poised to unlock transformative capabilities that will shape the future of technology and address pressing societal challenges.

Convergence of Nanotechnology with Other Technologies

Nanotechnology's convergence with other cutting-edge fields such as artificial intelligence (AI) and nanoelectronics is driving transformative advancements across various industries. This convergence leverages the unique properties of nanomaterials and nanostructures in order to enhance technological capabilities and to unlock novel applications.

Nanotech with Artificial Intelligence:
1. Synergistic Advancements:

The integration of nanotechnology and AI facilitates synergistic advancements by combining the precision of nanoscale manipulation with the cognitive abilities of AI algorithms. This collaboration enables the development of smart nanomaterials and nanodevices with enhanced functionalities and performance.

2. Materials Discovery:

AI algorithms analyse vast datasets in order to accelerate the discovery and design of nanomaterials with tailored properties for specific applications. This approach expedites the development of advanced materials for electronics, healthcare, energy, and beyond.

3. Predictive Modelling:

AI-powered simulations and modelling predict the behaviour of nanoscale systems, aiding in the optimization of manufacturing processes and the design of nanoelectronic devices. This predictive capability enhances efficiency and reduces time-to-market for nanotechnology-based products.

Nanoelectronics Integration:
1. Miniaturization:

Nanoelectronics involve the fabrication of electronic components at the nanoscale, leading to smaller, faster, and more energy-efficient devices. Integration with nanotechnology enables the development of nanoscale transistors, sensors, and memory devices with unprecedented performance.

2. Flexible Electronics:

Nanotechnology enables the integration of nanomaterials such as carbon nanotubes and graphene into flexible electronic circuits. These flexible electronics offer advantages such as lightweight, bendable, and wearable form factors, revolutionizing applications in healthcare, consumer electronics, and beyond.

3. Quantum Computing:

Nanoelectronics play a crucial role in the development of quantum computing platforms by providing the means to control and manipulate quantum states at the nanoscale. This integration holds promise for realizing ultra-powerful quantum computers capable of solving complex problems beyond the reach of classical computers.

In summary, the convergence of nanotechnology with AI and nanoelectronics represents a paradigm shift in technological innovation. By leveraging synergies between these fields, researchers and engineers are pioneering novel solutions with unprecedented capabilities, paving the way for a future defined by smarter, smaller, and more efficient technologies.

Nanotechnology for Environmental Remediation and Sustainable Nanomaterials

Nanotechnology offers promising solutions for environmental remediation through the development of innovative nanomaterials for pollution control and remediation. Additionally, there is a growing focus on sustainable nanomaterials to address environmental concerns and ensure eco-friendly practices in nanotechnology applications.

Nanomaterials for Pollution Control and Remediation:
1. Versatile Applications:

Nanomaterials, such as nanoparticles and nanocomposites, exhibit unique properties such as high surface area-to-volume ratio and reactivity, making them effective for various remediation techniques. These materials can adsorb, degrade, or immobilize contaminants in soil, water, and air.

2. In Situ and Ex Situ Remediation:

Nanoparticles are highly flexible for both in situ and ex situ remediation approaches. In situ remediation involves treating contaminants directly in the environment, while ex situ methods involve removing contaminated media for treatment elsewhere.

3. Successful Remediation Procedures:

Nanotechnology-based remediation procedures have shown promising results in treating various pollutants, including heavy metals, organic pollutants, and emerging contaminants. These procedures offer efficient, cost-effective, and sustainable alternatives to traditional remediation methods.

Sustainable Nanomaterials:

1. Green Nanotechnology:

The concept of green nanotechnology focuses on developing environmentally friendly nanomaterials and nanotechnology processes. This includes using renewable resources, reducing energy consumption, and minimizing waste generation throughout the nanomaterial life cycle.

2. Biodegradable Nanomaterials:

Future trends in sustainable nanomaterials emphasize the development of biodegradable nanomaterials that degrade naturally in the environment, reducing the risk of long-term environmental impact. These materials offer safer alternatives for various applications, including packaging, biomedical devices, and water treatment.

3. Renewable Nanomaterial Synthesis:

Researchers are exploring sustainable methods for synthesizing nanomaterials using renewable resources and environmentally benign processes. These approaches aim to reduce the reliance on hazardous chemicals and energy-intensive processes, promoting greener nanotechnology practices.

In summary, nanotechnology plays a crucial role in environmental remediation through the development of innovative nanomaterials for pollution control and remediation. Moreover, the shift towards sustainable nanomaterials reflects the growing awareness of environmental sustainability in nanotechnology research and applications.

Challenges and Ethical Considerations in Nanotechnology

Nanotechnology presents immense potential for scientific breakthroughs and technological advancements. However, along with its promise come various challenges and ethical considerations that must be addressed to ensure responsible development and deployment.

Ethical Considerations:
1. Risk Assessment:

Ethical concerns in nanotechnology involve the identification and assessment of potential hazards and risks associated with the use of nanomaterials and nanodevices. It's crucial to conduct thorough risk assessments to mitigate any adverse effects on human health and the environment.

2. Autonomy and Informed Consent:

The autonomy of individuals should be respected, ensuring that they have the necessary information to make informed decisions about the use of nanotechnology products or services. Transparency and clear communication regarding potential risks and benefits are essential.

3. Societal Implications:

Ethical considerations extend beyond individual autonomy to broader societal implications. Nanotechnology developments may have far-reaching social, economic, and political consequences that need to be carefully evaluated and addressed.

Overcoming Technological Challenges:
1. Scale-Up and Manufacturing:

One significant challenge in nanotechnology is scaling up laboratory-scale processes to commercial-scale manufacturing. Researchers are developing innovative manufacturing techniques and scalable processes in order to mass-produce nanomaterials and nanodevices efficiently.

2. Environmental Impact:

Nanomaterials and nanodevices may pose environmental risks if not properly managed. Efforts are underway to develop sustainable nanomaterial synthesis methods and implement effective waste management strategies in order to minimize environmental impact.

3. Regulatory Frameworks:

Establishing robust regulatory frameworks is essential to ensure the safe and responsible use of nanotechnology. Governments and regulatory bodies need to collaborate with researchers and industry stakeholders to develop and implement appropriate regulations that address safety, ethics, and environmental concerns.

4. Interdisciplinary Collaboration:

Overcoming technological challenges in nanotechnology requires interdisciplinary collaboration across various scientific disciplines, including physics, chemistry, biology, and engineering. By bringing together diverse expertise, researchers can address complex challenges more effectively.

In summary, addressing ethical considerations and overcoming technological challenges are essential for the responsible advancement of nanotechnology. By fostering interdisciplinary collaboration, implementing robust regulatory frameworks, and prioritizing ethical principles, we can harness the full potential of nanotechnology while ensuring its safe and sustainable integration into society.

Global Collaborations and Regulatory Frameworks in Nanotechnology

Nanotechnology, with its vast potential for scientific innovation and technological advancement, requires robust global collaborations and regulatory frameworks in order to ensure its safe and responsible development. Here's why global cooperation is crucial and an overview of the regulatory landscape:

Importance of Global Cooperation:
1. Knowledge Sharing:

Nanotechnology research often transcends geographical boundaries. Global collaborations facilitate the sharing of knowledge, expertise, and resources, accelerating scientific progress and innovation.

2. Addressing Emerging Challenges:

By pooling together diverse perspectives and capabilities from around the world, global cooperation enables proactive identification and mitigation of emerging challenges in nanotechnology research and application.

3. Harmonizing Standards:

Nanotechnology standards and regulations vary across countries. Global collaboration promotes the harmonization of standards, ensuring consistency in safety protocols, environmental protection measures, and ethical guidelines.

Regulatory Frameworks:
1. Current Landscape:

Nanotechnology regulatory frameworks are still evolving, often struggling to keep pace with rapid technological advancements. Many countries have established guidelines for the safe handling, production, and disposal of nanomaterials.

2. Anticipated Developments:

Future regulatory developments are expected to focus on enhancing safety assessment methodologies, improving risk management strategies, and strengthening international cooperation.

3. Risk-Based Approach:

Regulatory agencies are increasingly adopting a risk-based approach to nanotechnology regulation, prioritizing the assessment and management of potential risks associated with nanomaterial exposure to humans and the environment.

4. Interdisciplinary Collaboration:

Effective nanotechnology regulation requires interdisciplinary collaboration among scientists, policymakers, industry stakeholders, and regulatory agencies. Such collaboration fosters a comprehensive understanding of nanotechnology's implications and facilitates the development of targeted regulatory measures.

In summary, global collaborations and regulatory frameworks are indispensable for navigating the complexities of nanotechnology research and application. By fostering international cooperation, harmonizing standards, and adopting proactive regulatory approaches, the global community can harness the transformative potential of nanotechnology while ensuring its safe and ethical integration into society.

Interdisciplinary Applications of Nanotechnology and Societal Impacts

Nanotechnology's interdisciplinary nature facilitates its integration with various scientific fields, paving the way for groundbreaking applications and significant societal impacts.

Interdisciplinary Applications:
1. Medicine:

Nanotechnology holds immense promise in healthcare, enabling targeted drug delivery, early disease detection through biosensors, and precise imaging techniques. Collaborations between nanotechnology and biomedicine are driving innovations in personalized medicine.

2. Energy:

Nanotechnology contributes to renewable energy solutions by enhancing solar cell efficiency, developing advanced battery technologies, and improving energy storage systems. Collaborative efforts between nanotechnology and energy sciences aim to address global energy challenges sustainably.

3. Environmental Remediation:

Interdisciplinary research at the intersection of nanotechnology and environmental sciences focuses on developing nanomaterial-based solutions for pollution control, water purification, and remediation of contaminated sites. Such collaborations aim to mitigate environmental degradation and promote ecological sustainability.

Societal Impacts:
1. Healthcare Revolution:

Nanotechnology-enabled medical advancements offer personalized treatments, early disease detection, and minimally invasive procedures, improving healthcare outcomes and enhancing quality of life.

2. Clean Energy Adoption:

Nanotechnology's role in advancing renewable energy technologies accelerates the transition to clean energy sources, reducing reliance on fossil fuels, mitigating climate change, and promoting environmental sustainability.

3. Environmental Protection:

Nanotechnology-based solutions contribute to environmental conservation efforts by enabling efficient pollutant removal, sustainable water treatment, and eco-friendly manufacturing processes, safeguarding ecosystems and human health.

4. Enhanced Consumer Products:

Nanotechnology integration in consumer goods leads to the development of novel materials, coatings, and electronics with enhanced functionalities, durability, and performance, enriching everyday life experiences.

5. Economic Growth and Job Creation:

Nanotechnology-driven innovations stimulate economic growth by fostering entrepreneurship, attracting investments, and creating new job opportunities in research, development, and manufacturing sectors, contributing to socio-economic prosperity.

In essence, nanotechnology's interdisciplinary applications transcend disciplinary boundaries, offering transformative solutions to pressing societal challenges and ushering in a new era of innovation, sustainability, and improved quality of life.

Conclusion: Embracing the Future of Nanotechnology

In conclusion, the future of nanotechnology holds immense promise and potential for transformative advancements across various sectors. Emerging trends indicate a trajectory towards innovative solutions that will revolutionize industries, improve quality of life, and address pressing global challenges.

Summary of Future Prospects:
1. Healthcare Revolution:

Anticipated developments in nanomedicine include personalized treatments, targeted drug delivery systems, and advanced diagnostic tools, offering more effective and efficient healthcare solutions.

2. Sustainable Energy Solutions:

Nanotechnology is poised to enhance renewable energy technologies, such as solar cells and batteries, leading to increased efficiency, affordability, and widespread adoption of clean energy sources.

3. Environmental Remediation:

Nanomaterial-based solutions will play a crucial role in addressing environmental challenges, including pollution control, water purification, and remediation of contaminated sites, promoting sustainability and ecological preservation.

4. Advanced Materials and Manufacturing:

Nanotechnology will enable the development of novel materials with unprecedented properties, revolutionizing industries such as electronics, aerospace, and construction, and fostering innovation in manufacturing processes.

Call to Action:

As we stand on the cusp of a nanotechnological revolution, it is imperative to recognize the significance of continued research and exploration in this field. To fully realize the potential of nanotechnology and harness its benefits for society, we must:

1. Invest in Research:

Governments, academia, and industry stakeholders should allocate resources and funding towards fundamental and applied research in nanotechnology, supporting interdisciplinary collaborations and fostering innovation.

2. Promote Collaboration:

Encourage collaboration between scientists, engineers, policymakers, and stakeholders from diverse backgrounds to address complex challenges and maximize the societal impact of nanotechnological advancements.

3. Ensure Ethical and Responsible Development:

Prioritize ethical considerations, safety standards, and regulatory frameworks in order to guide the responsible development and deployment of nanotechnology, mitigating potential risks and safeguarding public health and the environment.

4. Educate and Engage:

Foster public awareness, education, and engagement initiatives in order to enhance understanding of nanotechnology's potential benefits and risks, empowering individuals to make informed decisions and participate in shaping its future.

By embracing these principles and taking proactive steps towards advancing nanotechnology, we can unlock unprecedented opportunities for innovation, sustainable development, and societal progress. Together, let us embark on a journey to explore the boundless possibilities of nanotechnology and build a brighter future for generations to come.

Dear Reader,

We hope you have found this book on nanotechnology to be both informative and inspiring. As you reach the end of your reading journey, we kindly ask you to take a moment to share your thoughts and experiences by writing a review.

Your review is essential in helping future readers understand the value and impact of this book. Here are some key points to consider while writing your review:

1. **Content and Coverage**: Briefly describe the scope of the book. What major topics does it cover? How comprehensively does it address various aspects of nanotechnology, from basic principles to advanced applications?

2. **Readability and Style**: Evaluate the writing style. Is the book written in a way that is easy to understand, even for those new to the field? Does it effectively balance technical detail with readability?

3. **Highlights and Key Insights**: Mention any particular sections or insights that stood out to you. Did the book offer any new perspectives or particularly innovative ideas?

4. **Author's Expertise**: Consider the author's background and how it influenced the content. Does the author bring a unique or authoritative perspective to the subject of nanotechnology?

5. **Practical Applications**: Reflect on how the book discusses the practical applications of nanotechnology. Are there specific examples or case studies that illustrate how nanotechnology is being used in real-world scenarios?

6. **Comparative Analysis**: If you have read other books on nanotechnology, how does this one compare? Does it offer a fresh perspective or fill gaps left by other works?

7. **Overall Impact**: Summarize your overall impression of the book. Would you recommend it to others interested in nanotechnology? Who would find this book most beneficial?

Your detailed and honest review will greatly assist others in deciding whether this book is the right choice for them. Moreover, it contributes to the broader conversation about nanotechnology, helping to advance understanding and interest in this crucial field.

Thank you for your time and effort in writing a review. Your feedback is highly valued.

Best regards,

Geoff Thomas PhD

Thank you for journeying with us through the fascinating world of nanotechnology. As we conclude this book, we are excited to announce that this is just the beginning. Our exploration of cutting-edge science and technology continues with 11 more books in this series. Each upcoming book delves into revolutionary topics shaping our future:

- **CRISPR**: Discover the groundbreaking gene-editing technology transforming genetics and medicine, allowing precise alterations to DNA for health and scientific advancements.
- **Artificial Intelligence**: Explore how AI is evolving, its capabilities to simulate human intelligence, and its impact on various industries and daily life.
- **Space**: Unveil the latest developments in space exploration, from NASA missions to private ventures, and how these advancements expand our understanding of the universe.
- **Neurotech**: Investigate technologies interfacing with the brain, enhancing cognitive functions, and offering new treatments for neurological disorders.
- **Quantum Computing**: Understand the principles of quantum mechanics applied to computing, promising unprecedented computational power and new problem-solving methods.
- **Precision Medicine**: Learn about personalized medical treatments tailored to individual genetic profiles, revolutionizing healthcare and treatment efficacy.
- **Renewable Energy**: Examine sustainable energy solutions, their innovations, and the global shift towards reducing carbon footprints.

- **Robotics**: Dive into advancements in robotics, their applications in various fields, and how they are changing the landscape of automation and human interaction.
- **Bioengineering**: Discover the engineering of biological systems for developing new technologies in medicine, agriculture, and environmental science.
- **Immunotherapy**: Explore cutting-edge treatments harnessing the immune system to fight diseases, offering new hope for conditions like cancer.
- **Telemedicine**: Investigate the rise of remote healthcare services, their technologies, and how they are making medical care more accessible and efficient.

Stay tuned for these enlightening explorations into the forefront of science and technology, each promising to expand your knowledge and imagination.

www.ingramcontent.com/pod-product-compliance
Lightning Source LLC
Chambersburg PA
CBHW071918210526
45479CB00002B/468